Springer Theses

Recognizing Outstanding Ph.D. Research

Aims and Scope

The series "Springer Theses" brings together a selection of the very best Ph.D. theses from around the world and across the physical sciences. Each volume has been selected for its scientific excellence by an internationally renowned institute or department. For greater accessibility to non-specialists, the published versions include an extended introduction, as well as a foreword by the student's supervisor explaining the special relevance of the work for the field. The series as a whole provides a valuable resource both for newcomers to the research fields, and for other scientists seeking detailed background information on special questions. Finally, it serves as an accredited documentation of the valuable contributions made by today's younger generation of scientists.

How to Apply

Theses prepared at internationally leading universities and institutes may be nominated for publication by supervisors with the endorsement of the head of department. Each such department is entitled to nominate one thesis per year. Enquirics and requests for a nomination form should be sent, together with the thesis, to Ms. Nathalie Berg (nathalie.berg@springer.com).

Eligibility Criteria

To qualify for consideration, theses should fulfil the following criteria:

- They must be written in good English.
- The topic should fall within the confines of Chemistry, Physics, Earth Sciences, Engineering and related interdisciplinary fields such as Materials, Nanoscience, Chemical Engineering, Complex Systems and Biophysics.
- The work reported in the thesis must represent a significant scientific advance.
- A maximum of 30% of the text may be a verbatim reproduction from the author's previous publications.
- They must have been examined and passed during the 12 months prior to nomination.
- The theses should have a clearly defined structure including an introduction accessible to new PhD students and scientists not expert in the relevant field.
- After acceptance, theses should be supplemented by a "Supervisor's Foreword" outlining the wider scientific relevance of the work.

Indexed by zbMATH.

Adam Michael William Powell

Magnetic Field Characterisation for Gravitational Free Fall Measurements of Antihydrogen in the ALPHA-g Experiment

Doctoral Thesis accepted by
the University of Calgary, Calgary, AB, Canada

Springer

Author
Dr. Adam Michael William Powell
Research Fellow, Experimental Physics
Department, CERN
Geneva
Switzerland

Supervisor
Dr. Timothy Friesen
Assistant Professor, Faculty of Science
Department of Physics and Astronomy
University of Calgary
Calgary, AB, Canada

ISSN 2190-5053 ISSN 2190-5061 (electronic)
Springer Theses
ISBN 978-3-032-07189-7 ISBN 978-3-032-07190-3 (eBook)
https://doi.org/10.1007/978-3-032-07190-3

This Springer imprint is published by the registered company Springer Nature Switzerland AG
The registered company address is: Gewerbestrasse 11, 6330 Cham, Switzerland

Supervisor's Foreword

Since its prediction and subsequent discovery almost 100 years ago, antimatter has fascinated both physicists and the general public alike. One of the questions that has long been speculated about is "Does antimatter fall down in the same way as matter does?" According to Einstein's General Relativity the acceleration of a body due to a gravitational field is independent of the nature of the body (the Weak Equivalence Principle or WEP). Antimatter should therefore fall down in precisely the same manner as matter. However, some have suggested that violation of the WEP may provide signs of quantum gravity and there have even been suggestions that antimatter could fall up (for which there are several compelling indirect arguments against). Even a tiny difference found in the gravitational behaviour of matter and antimatter would completely change our understanding of antimatter and gravity. Until the work reported here, there were no measurements of the motion of antimatter under the effect of gravity due to the difficulty producing and manipulating antimatter and the extreme weakness of gravity relative to the electromagnetic force.

Founded in 2005, the ALPHA collaboration at CERN has pioneered techniques to produce, trap, and studying antihydrogen atoms, the antimatter equivalent of hydrogen atoms. Antihydrogen atoms are ideal for testing the WEP with antimatter because they are neutral and not sensitive to stray electric fields. Antihydrogen is synthesized by ALPHA by combining antiprotons (provided by the Antiproton Decelerator facility at CERN) and positrons to create cold antihydrogen atoms that are confined in a magnetic minimum trap. ALPHA has gone on to demonstrate successful laser cooling and spectroscopy of antihydrogen to test the fundamental symmetry between antimatter and matter. Building on ALPHA's successes in producing and magnetically trapping antihydrogen atoms, ALPHA set out to build a new device, known as ALPHA-g, to probe the gravitational motion of antihydrogen.

Dr. Adam Powell's thesis presents the first ever measurement of the free-fall of antimatter using ALPHA-g and shows that's that antimatter does indeed fall down. The concept of the measurement is simple: release antihydrogen atoms from a vertical magnetic trap and measure the annihilation position of the anti-atoms as they leave the trap under the influence of gravity. However, carrying out the measurement is extremely challenging. A central challenge, which is the primary focus of Dr.

Powell's thesis, is the measurement and control of the magnetic fields that are used to trap and release the atoms. Dr. Powell developed precise magnetic field measurement techniques based on the measurement of the Electron Cyclotron Resonance (ECR) of electron clouds confined in the trap. He led an extensive campaign of ECR measurements to develop and characterize the magnetic fields and release protocol for antihydrogen release experiments. The thesis also describes key hardware contributions Dr. Powell made to improvements of particle traps in ALPHA and injection of microwave radiation to enable ECR measurements and future microwave spectroscopy in ALPHA-g. These key contributions, laid out clearly in Dr. Powell's excellent thesis, are just the first step of ALPHA's efforts to precisely measure the effect of gravity on antimatter to the 1% level and beyond and rigorously test the foundations of physics using antimatter.

Calgary, AB, Canada
September 2025

Timothy Friesen

Preface

The bound state of an antiproton and positron, antihydrogen, is an ideal test particle for comparisons between matter and antimatter as hydrogen has been studied extensively through history both experimentally and theoretically. The Antihydrogen Laser Physics Apparatus (ALPHA) collaboration has made significant progress on antihydrogen trapping, cooling, and spectroscopy in recent years. In a new apparatus, ALPHA-g, the collaboration aims to probe the effects of gravity on antimatter. This thesis will focus on the first measurement of the free-fall (or gravitational motion) of antimatter and the magnetic field measurements essential to this observation.

The ALPHA apparatus, particularly the ALPHA-g experiment, will be discussed in detail. This thesis will outline hardware improvements to the ALPHA-g's Penning trap as well as the microwave injection path. These developments were critical for the operation of ALPHA-g and electron cyclotron resonance (ECR) magnetometry. In addition, a design for a new Penning trap for ALPHA-g is shown that will be key in future measurements.

The electron cyclotron resonance (ECR) magnetometry technique used in ALPHA will be introduced. ECR is needed to measure, set, and monitor the magnetic fields in ALPHA-g, ECR is then the essential tool at the center of successful measurements using ALPHA-g. ECR was also used to constrain major systematic uncertainties through a host of applications of the technique in novel ways including fast repeating ECR and multiple simultaneous locations. This work is the first application of this technique in a high magnetic field gradient, which was an important obstacle to overcome in characterising the magnetic fields in ALPHA-g.

This breakthrough result of the first ALPHA-g measurement produced a ratio of antihydrogen gravitational acceleration and "normal gravity" of $a_{\bar{g}}/g = 0.75 \pm 0.13$ (statistical + systematic) $\pm\, 0.16$ (simulation). This ratio is consistent (within errors) with no difference between matter and antimatter gravity. This result has allowed ALPHA to rule out extreme difference between matter and antimatter such as no gravitational interaction or repulsive gravity.

Geneva, Switzerland — Adam Michael William Powell

Acknowledgements Taking a long time to finish a Ph.D. leads to the issue of having many people to thank for their years of support. The first person that has to be thanked in Dr. Tim Friesen. Joining UCalgary as Tim's first Ph.D. student has been an absolute privilege, he has been an excellent supervisor, mentor, and colleague as we have navigated the experience together.

Thank you to supervisory committee of Prof. Robert Thompson, Prof. Mike Wieser, and Prof. Brian Jackel for great comments and thought provoking questions over the years. Thank you to all members of the ALPHA-Canada faculty, particularly Prof. Makoto Fujiwara, for their continued support and trust.

Generally my ALPHA colleagues have been extraordinary. You have all taught me so much and pushed me to be better in all endeavours. The group of Ph.D.'s that started around the same time, Patrick, Asia, Jack, Peter, Mark, and Danielle all proved to be great friends and colleagues. A special thanks Dr. April Cridland, we survived the many months of building Penning traps together, you taught me a lot. Dr. Chris Rasmussen, thank you for sharing a wealth of knowledge and always being available to discuss the intricate details of many measurements for ALPHA-g. A thank you to Prof. Niels Madsen and Prof. Mike Charlton, without whom I would not have had the opportunity to join ALPHA, for continued guidance and support.

To my friends around Geneva and CERN, thank you for inviting me to things even though I said no I have to work 90% of the time, your friendship and caring has made it into a home. Particularly Alessandro, Jamie, Kostas, David, Eszter, Laurens, and Lily for bringing me into their little family. Luke Dykes for a shared enjoyment of wine and rugby. To Freya, thanks for keeping me sane during a difficult period around the COVID-19 pandemic and manic hours in the lab.

While I spent most of the time during this thesis at CERN there are some people in Calgary, Dr. Andrew Evans, Dr. Jake Flowerdew, and Eamon, who made my first winter in the real cold tolerable (despite my frequent complaints).

In Swansea, I could always count on a welcome home and a break from research with school, university, and SU friends. Lucy, thank you for supporting me through the final part of this long process. Sam and Ted, you guys bridge the gap between friends and family bringing both encouragement and a healthy dose of reality in good times and bad.

A huge thank you to my family (there are too many of you to name), you have always supported me even if most of you think my job is "he lives in Geneva". Finally, my parents. My mum, who has loved and supported me through life more than one can believe possible. My Dad, who always shown me that being a kind and decent person is the key to a happy life. Thank you both for your support, friendship, and generally being an outstanding human beings, this wouldn't have been possible without you.

Competing Interests The author has no competing interests to declare that are relevant to the content of this manuscript.

Contents

Symbols, Abbreviations, and Nomenclature

AD	Antiproton decelerator
AEGIS	Antihydrogen Experiment: Gravity, Interferometry, Spectroscopy
ALPHA	Antihydrogen laser physics apparatus
ALPHA-2	The spectroscopic measurement section of the ALPHA experiment
ALPHA-g	The gravity measurement section of the ALPHA experiment
ASACUSA	Atomic spectroscopy and collisions using slow antiprotons
ATHENA	Experiment at CERN producing the first cold antihydrogen
ATRAP	Antihydrogen Trap
BASE	Baryon antibaryon symmetry experiment
CCD	Charge-coupled device
CERN	Conseil Européen pour la Recherche Nucléaire
ECR	Electron cyclotron resonance
EEP	Einstein Equivalence Principle
ELENA	Extra low energy antiproton ring
FC	Faraday Cup
Fermilab	Fermi National Accelerator Laboratory
GBAR	Gravitational behaviour of antimatter at rest
GR	General Relativity
JPARC	Japan Proton Accelerator Complex
LEAR	Low Energy Antiproton Ring
MCP	Microchannel plate
NNP	Non-neutral plasma
PUMA	Antiproton unstable matter annihilations
rTPC	Radial Time Projection Chamber
RW	Rotating wall
SDREVC	Strong drive regime evaporative cooling
SiPM	Silicon Photo-multiplier
UCalgary	University of Calgary
WEP	Weak Equivalence Principle

Symbols

C	Charge conjugation
P	Parity inversion
T	Time reversal
$H/\bar{H}$	Hydrogen/Antihydrogen
$p/\bar{p}$	Proton/Antiproton
e^-/e^+	Electron/Positron
$\vec{B}$	Magnetic field
$\vec{E}$	Electric field
ω_c	Cyclotron frequency
ω_z	Axial bounce frequency
ω_m	Magnetron frequency
ω_+	Modified cyclotron frequency
T	Temperature
k_B	Boltzman constant
q	Particle electric charge
m_e	Mass of electron
m_H	Mass of (anti)hydrogen
λ_D	Debye length
B_A	Magnetic field at mirror A
B_G	Magnetic field at mirror G
z_A	Position of magnetic field maximum at mirror A
z_G	Position of magnetic field maximum at mirror G
$Bias$	The difference between magnetic field at the maxima of Mirrors A and G
g	Gravitational acceleration
$a_{\bar{g}}$	Local gravitational acceleration of antihydrogen

Chapter 1
Introduction

Every matter particle has an equivalent antimatter partner with the same properties but several inverted quantum numbers. Our current cosmological models suggest equal amounts of matter and antimatter were created in the moment of the Big Bang, therefore the imbalance of matter and antimatter in our observable universe now is a major puzzle in modern physics known as "the baryon asymmetry problem" [1–3]. Given this baryon asymmetry, the search for observable differences between matter and antimatter outside of current understanding is enticing. One might ask for example how does antimatter behave in the gravitational field of the Earth?

General relativity (GR) has been an incredibly successful theory of gravity but does have known limitations such as not accounting for dark matter and a lack of quantum gravity. The study of the effect of gravity on antimatter as a test of the weak equivalence principle (WEP) is then an important measurement in the progression towards a complete theory of gravity [4, 5]. Amongst theorists there is little expectation that there is a difference between the interaction of matter and antimatter with gravity, however there is also no compelling theoretical argument against a difference at the 1% level[6]. Some theorists even suggest antimatter having repulsive gravity is possible and the implications of such a result (along with arguments against) have been discussed elsewhere [7].

The study of antimatter particles has been ongoing since the 1930's [8, 9], however the direct observation of this gravity interaction has eluded physicists as the electromagnetic forces acting on charged particles are many orders of magnitude larger than any gravitational force on a single particle. This makes an electrically neutral test particle such as antihydrogen ($\bar{H}$), the bound state of a positron (e^+) and an antiproton ($\bar{p}$), a valuable tool in gravitational experiments with antimatter.

This thesis will show the work of the Antimatter Laser PHysics Apparatus (ALPHA) collaboration towards using antihydrogen to probe the motion of antimatter in Earth's gravitational field in the newly constructed ALPHA-g experiment. The ALPHA collaboration has a long history of groundbreaking results including the first trapping of antihydrogen [10], the first observation of a quantum transition in

A. M. W. Powell, *Magnetic Field Characterisation for Gravitational Free Fall Measurements of Antihydrogen in the ALPHA-g Experiment*, Springer Theses,
https://doi.org/10.1007/978-3-032-07190-3_1

antihydrogen [11], high precision spectroscopy [12], and laser cooling of antihydrogen atoms [13]. These developments have paved the way for the latest ground breaking results, the observation the effect of gravity on antimatter for the first time [14].

This introductory chapter will provide context for why the study of antimatter is of interest and give an overview of the history of antimatter research. I will then discuss the various experiments currently working towards gravity experiments with antihydrogen and their various techniques and challenges. The chapter will conclude with an outline of the remaining chapters in this thesis as well as a description of the authors contributions.

1.1 Equivalence Principles

In 1915 Einstein published the highly successful theory of general relativity (GR) [15] and its predictions have been consistently proven, beginning with the notable eclipse in 1919 [16] right up to numerous gravitational wave observations at LIGO and VIRGO of recent years [17]. A cornerstone to the thought process during the inception of GR is a set of guiding principles, Einstein equivalence principles (EEP) [18]:

1. The motion of a test particle is independent of mass or composition under the influence of gravity.
2. Gravitational fields couple to all mass in the same manner.
3. Local experiments cannot distinguish between free fall in gravity and uniform motion with no gravity.
4. An accelerated frame in special relativity is indistinguishable from a frame at rest in a gravitational field.

Statements 1 and 2 have been known since the time of Galileo and his free fall experiments and rely on the equivalence of inertial and gravitational mass. Statements 3 and 4 are visualized in the famous elevator thought experiments described by Einstein. Einstein proposed a thought experiment involving two elevators, one in a region of space with a low local gravitational field and moving at a constant motion, while the other is falling in Earth's gravitational field at $g \simeq 9.81\,\mathrm{m/s}$. A person in either elevator would experience a sense of weightlessness and therefore is unable to distinguish free fall under gravity from constant motion without gravity.

Even though the study of equivalence principles is hundreds of years old, the precision frontiers are still being explored today. The mass of an object can now be measured to incredibly high precision and provide tight constraints on these equivalence principles for matter. A torsion balance experiment can measure a torque that is produced from the interaction with an external field on two test masses hanging from a wire. This approach to measuring the effect of gravity is well established and results continue to confirm the equivalence of inertial and gravitational mass [19]. Another method for measuring free fall is using a matter-wave interferometer, this approach measures shifts in interference patterns caused by external fields [20, 21].

1.1.1 Weak Equivalence Principle

The weak equivalence principle (WEP) is a reduced version of the full EEP, removing local Lorentz invariance and local position invariance [18], and states that the inertial mass and gravitational mass of a particle are the same. The inertial mass is the mass of the particle which can interact with a force as given by Newton's second law while the gravitational mass is the mass interacting with a gravitational field. Due to Einstein's famous mass and energy equivalence, the gravitational mass is the sum of the various energies of a particle (rest mass, binding energy, strong and weak nuclear forces etc.) and any difference in that energy's coupling to gravity would appear as a difference between the inertial and gravitational masses.

1.2 CPT Symmetry

It was previously believed that under any one of the three operations, charge conjugation (C) parity inversion (P), or time reversal (T), all physics in the universe was invariant. These operations are as follows,

$$
\begin{aligned}
q &\xrightarrow{\mathrm{C}} -q \\
(x, y, z) &\xrightarrow{\mathrm{P}} (-x, -y, -z) \\
t &\xrightarrow{\mathrm{T}} -t
\end{aligned}
\tag{1.1}
$$

where q is electric charge, (x, y, z) are spatial coordinates, and t is time. Experiments in 1957 found the first evidence of P violations in beta decay experiments [22], while measurements of the neutrino helicity showed violations of both C and P [23]. Later theories suggested a combination of CP, i.e. performing both operations on a system together, was invariant while T symmetry was still accepted as stand-alone. CP symmetry violations in Kaon decays were discovered in 1964 [24] and a violation of T symmetry has been observed in the BaBar experiment looking at B mesons in 2012 [25].

Now the standard model assumes the invariance of a system under combined CPT operation [26] and there are a few experiments offering high-precision tests of CPT symmetry through direct comparison of matter and antimatter [27–31]. The existence of a CPT symmetry violation would open avenues into new physics with some of the possible ramifications discussed here [32], however, no violations have been discovered at this time.

1.3 Antimatter

1.3.1 Antimatter Gravity

There are however no constraints from CPT symmetry on the physics of antimatter in a matter gravitational field such as Earth's gravity. There have been some efforts to explore the consequences in cosmology of repulsive gravity [33–35].

Given the lack of constraints by CPT symmetry, the study of antimatter in Earth's gravity then becomes a test of the WEP outlined in Sect. 1.1.1. This thesis is focused on antimatter, however at the time Einstein was beginning to unravel the secrets of gravity with GR [15], the existence of antimatter had not yet been postulated. Antimatter is obviously then not explicitly included in GR, however, equivalence is expected to hold for matter and antimatter [7]. Using antihydrogen to probe this equivalence has long been proposed [4, 5] but unachievable experimentally until now.

1.3.2 History of Antimatter

The first theory hinting at the existence of antimatter was proposed by Paul Dirac in 1928 during his effort to include special relativity in quantum mechanics [36]. By reformulating the Schrödinger equation to satisfy relativity, Dirac's now famous equation,

$$(i\not{\partial} - m)\psi(x) = 0, \tag{1.2}$$

where m is a parameter interpreted as mass and $\not{\partial} = \gamma^{\mu}\delta_{\mu}$, gave negative energy solutions. The physical interpretation became that all spin $1/2$ particles must have a pair with an opposite charge, known as antimatter.

Just four years after the proposed existence of antimatter by Dirac, Carl David Anderson discovered the antimatter counterpart to the electron in a cloud chamber experiment, naming it the positron [8]. In this experiment, photographs of tracks in the cloud chamber made by cosmic rays were studied and it found tracks of the same curvature as an electron, but in opposite directions, indicating they had the same momentum but opposite electric charge. The next antimatter species observation came in the form of the antiproton in 1955 at the Bevatron in the University of California, Berkeley [37] through the collision of antiprotons with a copper target. Just a year later, the antineutron was observed in a charge exchange reaction at the same facility [38].

After establishing the existence of antimatter and confirming Dirac's predictions, a new problem arose as cosmological models became more developed. The models for the Big Bang suggest equal amounts of matter and antimatter should have been formed in the early universe. However, observations of the baryon-to-photon ratio in the universe suggest that a vast majority of the observable mass of the universe is in fact matter [3] and annihilation is the process of a matter/antimatter pair coming together and expelling energy in the form of photons and or charged particles. The question then arises, if equal amounts of antimatter and matter were produced in the early universe and they annihilate on contact, why do we have an excess of matter in the universe?

The Low Energy Antiproton Ring (LEAR) at the European Center for Nuclear Research (CERN) produced the first antihydrogen in 1995 [39]. Fermilab later reported the production of 99 antihydrogen atoms in their antiproton accumulator [40]. While the production of antihydrogen has been shown possible, the antihydrogen was formed in flight in these high-energy beam experiments, and thus were short lived before colliding with the apparatus. The study of properties and possible comparison with hydrogen requires a low-energy sample that can be stored for some time. This led to the installation of the Antiproton Decelerator at CERN providing antiprotons to experiments leading ATHENA [41] and ATRAP to produce low-energy antihydrogen by mixing low-energy antiprotons with positron plasma. The antihydrogen laser physics apparatus (ALPHA) became the first experiment to show trapping of antihydrogen atoms in a magnetic minimum [10] with ATRAP displaying similar results later [42].

Outside of fundamental physics, antimatter is a useful tool in many fields. The ability to measure coincident photons from positron annihilations has led to applications in medicine such as Positron Emission Tomography (PET) scans [43] and material science looking at imperfections through positron lifetimes [44].

1.4 Antimatter Experiments

The Antimatter Factory is the affectionate name given to the machinery at CERN that provides the only source of low-energy antiprotons in the world. Production of antiprotons begins with the acceleration of protons through the initial stages of the CERN accelerator complex starting with a linear accelerator before moving into synchrotrons. After this acceleration the protons, which are now at 26 GeV/c, are directed into an iridium target producing secondary particles including antiprotons [45].

Antiprotons are then magnetically filtered from the other products and steered into the antiproton decelerator (AD) at energies of 3.6 GeV/c [46]. As the name suggests the aim of the AD is to reduce the energy of the antiprotons from 3.6 GeV/c to 5.3 MeV/c. This is achieved through deceleration and cooling; the stochastic and electron cooling techniques were developed using the AD facility and were groundbreaking advances [47]. Until 2018 the antiprotons would then be distributed to the user experiments inside the antimatter factory in eight-hour alternating shifts. Since 2018 a new machine, the extra low energy antiproton ring (ELENA) [48] has been online and not only reduces the antiproton energy further to 100 keV/c but also provides 24-hour beam to four experiments at a time as of 2021 [49].

While the topic of this thesis will be the ALPHA experiment, there are a number of experiments within the facility as shown in Fig. 1.1 with a wide range of research

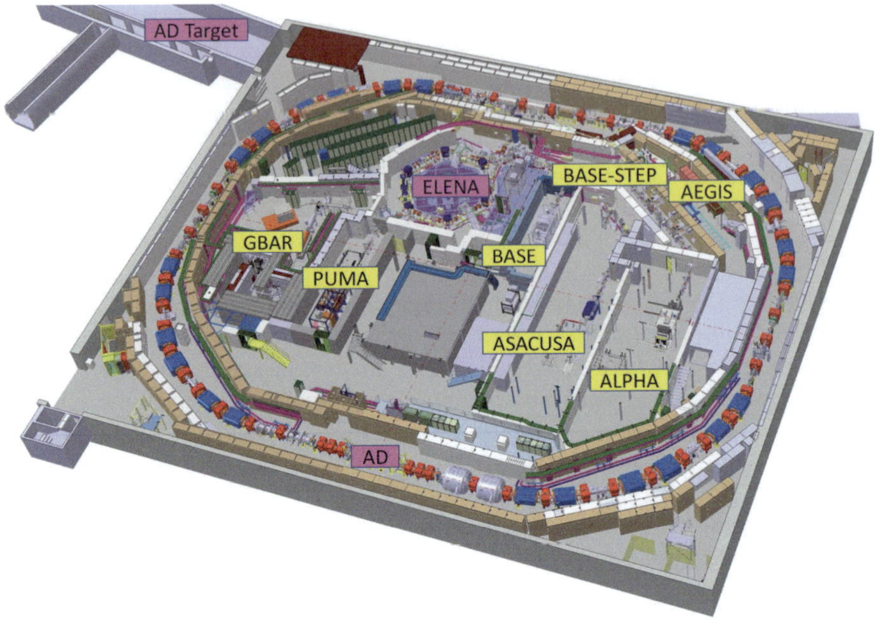

Fig. 1.1 An overview of the "Antimatter factory" at CERN showing decelerators and experimental areas. Used from publication with no modifications in accordance with open access permissions [49]

goals. Given the focus of this thesis will be on gravity experiments I will now only introduce all the experiments in the AD for reference, more detailed explanations for the gravity related experiments will be provided in Sect. 1.5 (Fig. 1.2).

- ASACUSA (Atomic Spectroscopy And Collisions Using Slow Antiprotons) is an experiment aiming to measure the ground state hyperfine splitting of antihydrogen in a cusp magnetic field [51] and performing spectroscopy of antiprotonic helium, a helium atom with antiprotons replacing an electron in an outer orbital [52].
- PUMA (antiProton Unstable Matter Annihilation) is a newly proposed and approved collaboration in the AD. PUMA proposes to use antiprotons to probe nuclear structure through interactions between antiprotons and heavy, unstable nuclei at the CERN ISOLDE facility [53]. This requires trapping antiprotons and transporting them across the CERN site while maintaining the confinement fields.
- BASE (Baryon Antibaryon Symmetry Experiment) measures proton and antiproton properties in the same Penning trap with non-destructive techniques.
- GBAR (Gravitational Behaviour of Antimatter at Rest) began work in 2018 to produce antihydrogen atoms at low temperatures, near rest, to study free fall [54]
- AEGIS (Antihydrogen Experiment: Gravity, Interferometry, Spectroscopy) aims to make an antihydrogen beam and study interference patterns produced by the antihydrogen interaction with external forces (such as gravity) [55].

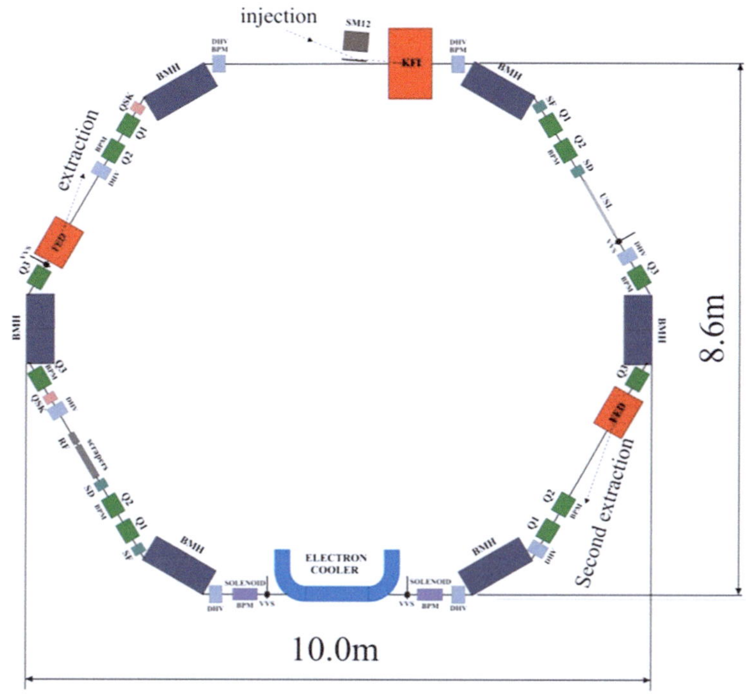

Fig. 1.2 A schematic of the extra low energy antiproton ring (ELENA) at CERN. Used from design report with no modifications in accordance with open access permissions [50]

Outside of the AD community, antimatter in the form of positrons and antimuons are available for study either as elementary particles in collision experiments, low energy studies in traps or in bound states. Positronium is the metastable bound state of an electron [56] and positron while muonium is the bound state of a positive muon and an electron [9]. Both of these exotic systems are hydrogen-like in energy states, but as neither has a nucleus there are no quantum chromodynamics (QCD) effects which have large theoretical uncertainties. These systems are therefore said to be an ideal test bed for quantum electrodynamics (QED), the theoretical description of how light and matter interact, through spectroscopy experiments [57, 58].

1.5 Antimatter Gravity Experiments

Table 1.1 offers a summary of the test particles or bound states and their discovery as laid out in Sect. 1.3.2. When selecting a test system for gravitational experiments with antimatter there are some important considerations. First, if one plans to measure the free fall of a particle in Earth's gravitational field it is clear a charged particle is not suitable as forces from electromagnetic fields are overwhelmingly greater than

Table 1.1 Summary of antimatter test particles/systems discovered

Particle	Symbol	Discovery (year)	Electric charge (e)	Lifetime	References
Positron	e^+	1932	+1	Stable	[8]
Positive muon	μ^+	1936	+1	2.2μ s	[9]
Antiproton	$\bar{p}$	1955	−1	Stable	[37]
Antineutron	$\bar{n}$	1956	0	Unstable (mins)	[38]
Para-positronium	pPs	1951	0	0.1244 ns	[56]
Ortho-positronium	oPs	1951	0	138.6 ns	[56]
Muonium	Mu	1960	0	2.2 μ s	[59]
Antihydrogen	$\bar{H}$	1995	0	Stable	[39]

any effect of gravity making experiments challenging [60]. The Baryon Antibaryon Symmetry Experiment (BASE) performs precision measurements of antiprotons and protons and extract gravitational information without attempting to measure a free fall. Comparison of the magnetic moment in protons and antiprotons are carried out by probing the Larmor and cyclotron frequencies of each species in a Penning trap. Extracting the charge to mass ratio of protons and antiprotons then allows a comparison of the mass of a proton and antiproton. If a difference between the frequencies of a proton/antiproton is observed this could indicate a different coupling of antimatter to gravitational fields [61], however this approach has been criticised for relying on an absolute gravitational potential [62]. To provide constraints to this difference in coupling the BASE experiment repeated their measurement campaign throughout the Earth/Sun orbit to average the gravitational potential [63].

As previously discussed, neutral particles or bound states are appealing for free fall experiments. Antineutrons are not ready available, especially at low enough energy to observe gravitational effects and are unstable. Antihydrogen is the ideal combination of electrically neutral and stable, several experiments are working towards a free fall measurement.

A relatively new experiment to the AD community, GBAR (Gravitational behaviour of antimatter at rest), has similar objectives to ALPHA but with different methods. As the name suggests this experiment requires extremely low energy antihydrogen to claim an understanding of "at rest" behaviour. The intention is to produce antihydrogen ions in collisions between antiprotons and positronium such that the antihydrogen ions are still trappable in charged particle traps. Cooling methods such as sympathetic cooling with laser-cooled ions of other species (beryllium for example [64]) can then be implemented. Once the antihydrogen ions have been cooled sufficiently, the extra positron can be photoionised leaving a no longer confined, at "rest" antihydrogen atom. This method is exciting and has the potential for high precision measurements, however the low cross section of antihydrogen ion production requires a high flux of positronium (and therefore positrons) as well as very low energy antiprotons, meaning there are several developments required before results will be possible. Progress is being made in dealing with the issues and exciting results, including the first production of antihydrogen atoms by GBAR [54].

The AEGIS collaboration aims to measure the vertical deflection of a horizontally propagating antihydrogen beam due to Earth's gravity. Using a pulsed antihydrogen beam and a Morié deflectometer, the interference pattern of antihydrogen can give a precise measurement of gravitational acceleration [55, 65]. The rate of antihydrogen production in this scheme is a current limiting factor.

Mixed matter and antimatter systems that are electrically neutral provide another possible test system for gravity's interaction with antimatter. Muonium gravity experiments with similar techniques are also in development using ultra cold muonium beams at the JPARC facility [59]. The interesting motivation for measurements with the muonium system is that a much larger fraction of the mass of the atom is antimatter compared to others discussed here. A free fall measurement of positronium is an ambition of some groups and has long been discussed [66]. To allow a long enough time of flight to observe free fall the positronium must be in a high Rydberg state due to the systems short lifetime, some advances have been made towards this goal [58].

The latest experiment added to the list is the topic of this thesis, ALPHA-g, is a new apparatus within the ALPHA setup. ALPHA-g takes the existing ALPHA production methods of antihydrogen and moves to a vertical trapping system, allowing for measurements that are sensitive to gravitational potential differences across the trap. I will discuss all the details in the coming chapters as well as the first results published in 2023 [14] but will provide a brief description here for context. ALPHA produces antihydrogen by the slow merging of low temperature positron and antiproton plasmas confined in a Penning-Malmberg trap. The formed antihydrogen is confined in a magnetic minimum trap produced by an octupole and mirror coils around the Penning trap electrodes. As previously mentioned this magnetic trap in ALPHA-g is vertical and as such has a gravitational potential difference across the trap that adds a linear term to the trapping potential. If the magnetic field produced by the mirror coils is slowly removed, the antihydrogen will escape in a direction biased by gravity and annihilate on the walls of the apparatus. ALPHA-g uses an annihilation detector to study the position of annihilations of escaping antihydrogen the vertical confining field is removed to extract information about gravity.

1.6 Comparison of Gravity Measurement Techniques with Antihydrogen

Now that a brief description of the ALPHA-g methodology has been given, it is prudent to examine the strengths and weaknesses of this method versus other previously discussed experiments (Table 1.2).

ALPHA-g has the benefit of a proven mechanism for the production and confinement of antihydrogen. It is limited by the ability to control and measure magnetic

Table 1.2 Comparison of antihydrogen gravity experiments

	ALPHA	AEGIS	GBAR
Formation method	Three body recombination	Charge exchange	Charge exchange (twice)
Current production rate	10 per AD cycle	–	1 per AD cycle ($\bar{H}$)
Experiment type	Free fall	Interferometery	Free fall
Target precision	1% of g	1% of g	1% of g
# of particles required	$\mathcal{O}(10^4)\bar{H}$	$\mathcal{O}(10^6)\bar{H}$	

fields inside the trapping regions. The addition of laser-cooled antihydrogen and the use of a region of the apparatus designed to minimise uncontrolled magnetic fields make the aim of a 1% precision determination of gravitational acceleration achievable. The details of this will be discussed in later chapters once the technical details of ALPHA-g have been fully explored.

The GBAR experiment aims to produce very low energy antihydrogen ions and then release as antihydrogen atoms by photoionisation. This has the benefit of not using magnetic fields in trapping antihydrogen which has been shown to exert a force on the atoms. The free fall chamber can be shielded to reach very low, stable background magnetic fields and will require field mapping.

The AEGIS experiment produces a low energy beam of antihydrogen atoms by the charge exchange mechanism shown in Sect. 2.21. This method has so far produced a low yield of antihydrogen atoms. Again this method has the advantage of not trapping the atoms in a magnetic field and so can use a shielded region for it's deflection measurement. This also comes with a field mapping requirement. The deflection method itself uses a light interferometer for calibration.

1.7 Thesis Outline

This thesis is the culmination of extensive efforts to directly probe the motion of antimatter in Earth's gravitational field for the first time. This introductory chapter provides motivation for studying antimatter systems generally and gives an overview of the antimatter experimental community. I will give background on trapping of charged particles using Penning-Malmberg traps and neutral atom confinement in a magnetic minima in Chap. 2. I will outline the ALPHA apparatus in Chap. 3 and hardware upgrades made for ALPHA-g will offer the reader an idea of the complexity of the experiment and limitations in Chap. 4.

There are three results chapters that will highlight my contributions to the ALPHA collaboration during this PhD research. Chapter 5 will show development work on magnetic field measurements essential to a successful determination of gravity's effect on antihydrogen. Chapter 6 will show the results from the first release experiments in ALPHA-g. This work will conclude with Chap. 7 showing the magnetic field measurements used during the release experiments and auxiliary measurements to constrain systematic uncertainties.

1.8 Author Contributions

Due to the nature of the ALPHA collaboration there is always overlap between collaborators in work carried out. The scale of ALPHA-g and the time frame for design, construction, assembly, and performing a free fall experiment lead to multiple people working closely on projects.

For work shown in Chap. 4, design and testing of Penning trap improvements were in collaboration with Dr. April Cridland. Design of the microwave injection system was led by Prof. Daniel Miranda Silveira with significant design input from a group including myself, Prof. Mike Hayden, Dr. William Bertche, Dr. Chris Ørum Rasmussen, and Dr. Timothy Friesen.

Development of the ECR technique in Chap. 5 built on the work of Dr. Eric Hunter, who implemented the reservoir ECR technique in ALPHA after its conception at the University of California Berkeley. The focus of this chapter is to introduce the technique, introduce developments to the technique necessary for ALPHA-g, and show the novel measurements in a high magnetic field gradient and simultaneously in multiple locations.

Chapter 6 shows the first controlled magnetic release of antihydrogen for gravitational studies. While this work was heavily collaborative, I led the development of the experimental protocol for release studies. I produced the first antihydrogen in ALPHA-g (both trapped and untrapped) and produced the first experimental escape curve.

Chapter 7 provides the details of a magnetic field characterisation campaign using ECR I led with input from Dr. Chris Ørum Rasmussen, Andrew Christensen, and other members of the collaboration. I developed the ECR techniques used to constrain the release experiment uncertainties. Data was taken across many weeks by a number of members of the collaboration as the experiment was operating 24 h a day.

References

1. Riotto A (1998) Theories of baryogenesis. In: ICTP summer school in high-energy physics and cosmology, pp 326–436
2. Sather E (1996) The mystery of the Matter Asymmetry. SLAC Beam Line 26:31–37
3. Bertone G, Hooper D, Silk J (2005) Particle dark matter: evidence, candidates and constraints. Phys Rep 405(5):279–390. ISBN: 0370-1573. https://doi.org/10.1016/j.physrep.2004.08.031. https://www.sciencedirect.com/science/article/pii/S0370157304003515

4. Hughes RJ (1993) Fundamental symmetry tests with antihydrogen. Nucl Phys A 558:605–624. ISBN: 0375-9474. https://doi.org/10.1016/0375-9474(93)90423-U. https://www.sciencedirect.com/science/article/pii/037594749390423U
5. Brown B et al (1997) Comparing the antiproton and proton and progress toward cold antihydrogen. Nucl Phys B Proc Suppl 56(1). In: Proceedings of the fourth biennial conference on low energy antiproton physics, pp 326–335. ISBN: 0920-5632. https://doi.org/10.1016/S0920-5632(97)00295-8. https://www.sciencedirect.com/science/article/pii/S0920563297002958
6. Fischler M, Lykken J, Roberts T (2008) Direct observation limits on antimatter gravitation. arXiv:0808.3929 [hep-th]
7. Nieto MM, Goldman T (1991) The arguments against "antigravity" and the gravitational acceleration of antimatter. Phys Rep 205.5, pp. 221–281. ISBN: 0370-1573. https://doi.org/10.1016/0370-1573(91)90138-C. https://www.sciencedirect.com/science/article/pii/037015739190138C
8. Anderson CD (1933) The Positive Electron. Phys Rev 43:491–494. https://doi.org/10.1103/PhysRev.43.491
9. Hughes VW et al (1960) Formation of Muonium and Observation of its Larmor Precession. Phys Rev Lett 5:63–65. https://doi.org/10.1103/PhysRevLett.5.63
10. Andresen GB et al (2010) Trapped antihydrogen. In: *Nature* 468 , pp. 673–676. ISBN: 1476-4687. https://doi.org/10.1038/nature09610
11. Amole C et al (2012) Resonant quantum transitions in trapped antihydrogen atoms. In: *Nature* 483, pp. 439–443. ISBN: 1476-4687. https://doi.org/10.1038/nature10942
12. Ahmadi M et al (2018) Characterization of the 1S–2S transition in antihydrogen. In: *Nature* 557, pp. 71–75. ISBN: 1476-4687. https://doi.org/10.1038/s41586-018-0017-2
13. Baker CJ et al (2021) Laser cooling of antihydrogen atoms. In: *Nature* 592, pp. 35–42. ISBN: 1476-4687. https://doi.org/10.1038/s41586-021-03289-6
14. Anderson EK et al (2023) Observation of the effect of gravity on the motion of antimatter. In: *Nature* 621, pp. 716–722. ISBN: 1476-4687. https://doi.org/10.1038/s41586-023-06527-1
15. Einstein A (1915) Fundamental ideas of the general theory of relativity and the application of this theory in astronomy. In: *Preussische Akademie der Wissenschaften, Sitzungsberichte* 315
16. Dyson FW, Eddington AS, Davidson C (1920) IX. A determination of the deflection of light by the sun's gravitational field, from observations made at the total eclipse of May 29, 1919. In: *Philosophical Transactions of the Royal Society of London. Series A, Containing Papers of a Mathematical or Physical Character* 220.571-581, pp. 291–333. https://doi.org/10.1098/rsta.1920.0009
17. Olsen S et al (2022) New binary black hole mergers in the LIGO-Virgo O3a data. In: *Phys. Rev. D* 106, p. 043009. https://doi.org/10.1103/PhysRevD.106.043009
18. Haugan MP, Lämmerzahl C, Principles of Equivalence: Their Role in Gravitation Physics and Experiments That Test Them. In: *Lecture Notes in Physics*. Springer Berlin Heidelberg, pp. 195–212. ISBN: 9783540412366. https://doi.org/10.1007/3-540-40988-2_10
19. Touboul P et al (2022) $MICROSCOPE$ Mission: Final Results of the Test of the Equivalence Principle. Phys Rev Lett 129:121102. https://doi.org/10.1103/PhysRevLett.129.121102
20. Hohensee MA et al (2011) Equivalence Principle and Gravitational Redshift. Phys Rev Lett 106:151102. https://doi.org/10.1103/PhysRevLett.106.151102
21. Wolf P et al (2010) Atom gravimeters and gravitational redshift. In: *Nature* 467.7311. ISBN: 1476-4687
22. Wu CS et al (1957) Experimental Test of Parity Conservation in Beta Decay. Phys Rev 105:1413–1415. https://doi.org/10.1103/PhysRev.105.1413
23. Goldhaber M, Grodzins L, Sunyar AW (1958) Helicity of Neutrinos. Phys Rev 109:1015–1017. https://doi.org/10.1103/PhysRev.109.1015
24. Christenson JH et al (1964) Evidence for the 2π Decay of the K_2^0 Meson. Phys Rev Lett 13:138–140. https://doi.org/10.1103/PhysRevLett.13.138
25. Lees JP et al (2012) Observation of Time-Reversal Violation in the B^0 Meson System. Phys Rev Lett 109:211801. https://doi.org/10.1103/PhysRevLett.109.211801

26. Lüders G (1957) Proof of the TCP theorem. In: *Annals of Physics* 2.1, pp. 1–15. ISBN: 0003-4916. https://doi.org/10.1016/0003-4916(57)90032-5. https://www.sciencedirect.com/science/article/pii/0003491657900325
27. Miller JP, de Rafael E, Roberts BL (2007) Muon (g 2): experiment and theory. In: *Reports on Progress in Physics* 70.5, R03. https://doi.org/10.1088/0034-4885/70/5/R03
28. DiSciacca J et al (2013) One-particle measurement of the antiproton magnetic moment. Phys Rev Lett 110:130801. https://doi.org/10.1103/PhysRevLett.110.130801
29. Schwingenheuer B et al (1995) CPT tests in the neutral kaon system. Phys Rev Lett 74:4376–4379. https://doi.org/10.1103/PhysRevLett.74.4376
30. Van Dyck RS, Schwinberg PB, Dehmelt HG (1987) New high-precision comparison of electron and positron g factors. Phys Rev Lett 59:26–29. https://doi.org/10.1103/PhysRevLett.59.26
31. Gabrielse G et al (1999) Precision Mass Spectroscopy of the Antiproton and Proton Using Simultaneously Trapped Particles. Phys Rev Lett 82:3198–3201. https://doi.org/10.1103/PhysRevLett.82.3198
32. Charlton M, Eriksson S, Shore GM (2020) Antihydrogen and Fundamental Physics. Springer International Publishing, Cham, Switzerland. ISBN: 978-3-030-51713-7. https://doi.org/10.1007/978-3-030-51713-7
33. Villata M (2011) CPT symmetry and antimatter gravity in general relativity. In: *Europhys. Lett.* 94.2, p. 20001. ISBN: 0295-5075. https://doi.org/10.1209/0295-5075/94/20001
34. Dimopoulos C et al (2023) Hubble law and acceleration curve emerges in a repulsive matter-anti matter galaxies simulations. In: *Astroparticle Physics* 147, p. 102806. ISBN: 0927-6505. https://doi.org/10.1016/j.astropartphys.2022.102806. https://www.sciencedirect.com/science/article/pii/S0927650522001074
35. Hajdukovic DS (2012) Quantum vacuum and virtual gravitational dipoles: the solution to the dark energy problem? In: *Astrophys. Space Sci.* 339.1, pp. 1–5. ISBN: 1572-946X. https://doi.org/10.1007/s10509-012-0992-y
36. Dirac PAM, Fowler RH (1928) The quantum theory of the electron. In: *Proceedings of the Royal Society of London. Series A, Containing Papers of a Mathematical and Physical Character* 117.778, pp. 610–624. https://doi.org/10.1098/rspa.1928.0023
37. Chamberlain O et al (1955) Observation of Antiprotons. Phys Rev 100:947–950. https://doi.org/10.1103/PhysRev.100.947
38. Cork B et al (1956) Antineutrons Produced from Antiprotons in Charge-Exchange Collisions. Phys Rev 104:1193–1197. https://doi.org/10.1103/PhysRev.104.1193
39. Baur G et al (1996) Production of antihydrogen. In: *Physics Letters B* 368.3, pp. 251–258. ISBN: 0370-2693. https://doi.org/10.1016/0370-2693(96)00005-6. https://www.sciencedirect.com/science/article/pii/0370269396000056
40. Blanford G et al (1998) Observation of Atomic Antihydrogen. Phys Rev Lett 80:3037–3040. https://doi.org/10.1103/PhysRevLett.80.3037
41. Amoretti M et al (2002) Production and detection of cold antihydrogen atoms. In: *Nature* 419, pp. 456–459. ISBN: 1476-4687. https://doi.org/10.1038/nature01096
42. Gabrielse G et al (2012) Trapped Antihydrogen in Its Ground State. Phys Rev Lett 108:113002. https://doi.org/10.1103/PhysRevLett.108.113002
43. Humm JL, Rosenfeld A, Del Guerra A (2003) From PET detectors to PET scanners. Eur J Nucl Med Mol Imaging 30:1574–1597
44. Siegel RW (1980) Positron annihilation spectroscopy. In: *Annual Review of Materials Science* 10.1, pp. 393–425
45. Boimond D et al (1994) Consolidation of the 400 kA magnetic horn for AAC antiproton production. Tech. rep, CERN
46. Maury S (1997) The antiproton decelerator: AD. In: *Hyperfine Interactions* 109.1, pp. 43–52
47. Bell M et al (1979) Electron cooling experiment at CERN. In: *Physics Letters B* 87.3, pp. 275–280. ISBN: 0370-2693. https://doi.org/10.1016/0370-2693(79)90982-1. https://www.sciencedirect.com/science/article/pii/0370269379909821
48. Maury S et al (2014) ELENA: the extra low energy anti-proton facility at CERN. In: *Hyperfine Interact.* 229.1, pp. 105–115. ISBN: 1572-9540. https://doi.org/10.1007/s10751-014-1067-y

49. Carli C et al (2022) ELENA: Bright Perspectives for Low Energy Antiproton Physics. In: *Nuclear Physics News* 32.3 , pp. 21–27. https://doi.org/10.1080/10619127.2022.2100646
50. edited by Chohan V (2014) *Extra Low ENergy Antiproton ring (ELENA) and its Transfer Lines*. [Online; accessed 16. Sep. 2024]. https://doi.org/10.5170/CERN2014002. https://cds.cern.ch/record/1694484/files/CERN-2014-002.pdf
51. Malbrunot C et al (2018) The ASACUSA antihydrogen and hydrogen program: results and prospects. In: *Philosophical Transactions of the Royal Society A: Mathematical, Physical and Engineering Sciences* 376.2116, p. 20170273. https://doi.org/10.1098/rsta.2017.0273
52. Sótér A et al (2022) High-resolution laser resonances of antiprotonic helium in superfluid 4He. In: *Nature* 603, pp. 411–415. ISBN: 1476-4687. https://doi.org/10.1038/s41586-022-04440-7
53. Aumann T et al (2022) PUMA, antiProton unstable matter annihilation. In: *Eur. Phys. J. A* 58.5, pp. 1–69. ISBN: 1434-601X. https://doi.org/10.1140/epja/s10050-022-00713-x
54. Adrich P et al (2023) Production of antihydrogen atoms by 6 keV antiprotons through a positronium cloud. In: arXiv:2306.15801 [hep-ex]
55. Aghion S et al (2014) A moiré deflectometer for antimatter. In: *Nat. Commun.* 5.4538, pp. 1–6. ISBN: 2041-1723. https://doi.org/10.1038/ncomms5538
56. Deutsch M (1951) Evidence for the Formation of Positronium in Gases. Phys Rev 82:455–456. https://doi.org/10.1103/PhysRev.82.455
57. Kanda S et al (2021) New precise spectroscopy of the hyperfine structure in muonium with a high-intensity pulsed muon beam. In: *Physics Letters B* 815, p. 136154. ISBN: 0370-2693. https://doi.org/10.1016/j.physletb.2021.136154. https://www.sciencedirect.com/science/article/pii/S0370269321000940
58. Adkins GS, Cassidy DB, Pérez-Ríos J (2022) Precision spectroscopy of positronium: Testing bound-state QED theory and the search for physics beyond the Standard Model. Phys Rep 975. Precision spectroscopy of positronium: testing bound-state QED theory and the search for physics beyond the Standard Model, pp. 1–61. ISBN: 0370-1573. https://doi.org/10.1016/j.physrep.2022.05.002. https://www.sciencedirect.com/science/article/pii/S0370157322001508
59. Antognini A et al (2018) Studying Antimatter Gravity with Muonium. In: *Atoms* 6.2. ISBN: 2218-2004. https://doi.org/10.3390/atoms6020017. https://www.mdpi.com/2218-2004/6/2/17
60. Witteborn FC, Fairbank WM (1968) Experiments to determine the Force of Gravity on Single Electrons and Positrons. In: *Nature* 220, pp. 436–440. ISBN: 1476-4687. https://doi.org/10.1038/220436a0
61. Hughes RJ (1990) Constraints on new macroscopic forces from gravitational redshift experiments. Phys Rev D 41:2367–2373. https://doi.org/10.1103/PhysRevD.41.2367
62. Chardin G, Manfredi G (2018) Gravity, antimatter and the Dirac-Milne universe. In: *Hyperfine Interact.* 239.1, pp. 1–13. ISBN: 1572-9540. https://doi.org/10.1007/s10751-018-1521-3
63. Borchert MJ et al (2022) A 16-parts-per-trillion measurement of the antiproton-to-proton charge–mass ratio. In: *Nature* 601, pp. 53–57. ISBN: 1476-4687. https://doi.org/10.1038/s41586-021-04203-w
64. Baker CJ et al (2021) Sympathetic cooling of positrons to cryogenic temperatures for antihydrogen production. In: *Nat. Commun.* 12.6139, pp. 1–8. ISBN: 2041-1723. https://doi.org/10.1038/s41467-021-26086-1
65. Amsler C et al (2021) Pulsed production of antihydrogen. In: *Commun. Phys.* 4.19, pp. 1–11. ISBN: 2399-3650. https://doi.org/10.1038/s42005-020-00494-z
66. Mills AP, Leventhal M (2002) Can we measure the gravitational free fall of cold Rydberg state positronium? In: *Nuclear Instruments and Methods in Physics Research Section B: Beam Interactions with Materials and Atoms* 192.1, pp. 102–106. ISBN: 0168-583X. https://doi.org/10.1016/S0168-583X(02)00789-9. https://www.sciencedirect.com/science/article/pii/S0168583X02007899

Chapter 2
Antihydrogen Production and Trapping in ALPHA

Now that opening chapter has provided appropriate motivation for carrying out gravitational measurements with antihydrogen has been provided, this chapter will give a brief overview of antihydrogen production and trapping in ALPHA (which has been extensively documented in other PhD thesis' [1, 2]). This will begin with details of charged particle trapping using a Penning trap in Sect. 2.1, before moving on to discuss properties of non-neutral plasmas and their confinement in a Penning trap in Sect. 2.2. Once the basics of charged particle confinement and collective behaviours have been laid out, the details of the plasma manipulations and diagnostics available can be shown in Sect. 2.3. In Sect. 2.4 I will discuss various mechanisms for formation of antihydrogen before Sect. 2.5 where I will show the confinement method used in ALPHA to produce a population of trapped antihydrogen ready for experimentation.

2.1 Penning Traps

The production of antihydrogen begins with confinement of the constituent charged particles, antiprotons and positrons. The Penning trap has a long history in experiments with charged particles from precision measurements with single particles to non-neutral plasmas to mass spectrometery. A Penning trap combines a strong uniform magnetic field with a static quadrupole electric field produced by hyperbolic electrodes for three dimensional confinement of a charged particles motion. Due to the Lorentz force, a single particle in a Penning trap has the equation of motion

$$\vec{F} = m\frac{d^2\vec{r}}{dt^2} = q(-\vec{\nabla}\phi + \vec{v} \times \vec{B}). \tag{2.1}$$

A. M. W. Powell, *Magnetic Field Characterisation for Gravitational Free Fall Measurements of Antihydrogen in the ALPHA-g Experiment*, Springer Theses,
https://doi.org/10.1007/978-3-032-07190-3_2

If the magnetic field is produced by a solenoid and is along the z axis,

$$\vec{B} = B_0\hat{z}, \tag{2.2}$$

with an electric potential $\phi(x, y, z)$ defined as,

$$\phi(x, y, z) = \frac{V_0(2z^2 - x^2 - y^2)}{R^2}, \tag{2.3}$$

where R is a parameter determined by the electrode geometry and V_0 is the potential applied to the trap electrodes. Simply inserting Eqs. 2.2 and 2.3 into the equation of motion and studying each component one finds,

$$\frac{d^2x}{dt^2} = \frac{q}{m}\left(\frac{2V_0x}{R^2} - B_0\frac{dy}{dt}\right), \tag{2.4}$$

$$\frac{d^2y}{dt^2} - \frac{q}{m}\left(\frac{2V_0y}{R^2} - B_0\frac{dx}{dt}\right), \tag{2.5}$$

$$\frac{d^2z}{dt^2} = -\frac{4qV_0z}{mR^2} = \omega_z^2 z. \tag{2.6}$$

It is then clear that the z motion of the particle has no magnetic field dependence and carries out a simple harmonic motion at the frequency ω_z, which will be referred to as the axial bounce frequency.

To deal with the x and y components in Eqs. 2.4 and 2.5, we define $u = x + iy$ and the equation of motion becomes

$$\frac{d^2u}{dt^2} - i\omega_c\frac{du}{dt} - \frac{\omega_z u}{2} = 0. \tag{2.7}$$

This has solutions $\omega_\pm = \omega_c/2 \pm \omega_1$, where $\omega_c = qB/m$ is the cyclotron frequency and $\omega_1 = \sqrt{\omega_c^2/4 - \omega_z^2/2}$. The high frequency solution ω_+ is the adjusted cyclotron frequency. The low frequency solution, ω_- is equivalent magnetron frequency ω_m resulting from the $\vec{E} \times \vec{B}$ drift. A number of variations of the Penning trap have been developed for different applications but the basic premise remains the same. One such variation, known as the Penning-Malmberg trap, is implemented by ALPHA and uses cylindrical electrodes to provide an open ended trap allowing particle transfers and access. While cylindrical electrodes do not produce a perfect quadrupole field, particles can still be trapped with some adjustments to the motional frequencies. There are four Penning-Malmberg traps in ALPHA for different applications including antiproton catching and positron accumulation. I will discuss the particular details and design considerations of the ALPHA Penning traps that are used to produce antihydrogen in Chap. 4 along with recent improvements specifically for ALPHA-g.

Table 2.1 Approximate values for different motion frequencies in electron/positron, and antiproton plasmas in ALPHA assuming a 1 T field

Parameter	e^-/e^+	Antiprotons
f_c	28 GHz	15 MHz
f_z	20 MHz	500 kHz
f_m	5 kHz	5 kHz

Approximate values for the various oscillation frequencies for electrons/positrons and antiprotons in ALPHA Penning traps at 1 T can be seen in Table 2.1.

2.2 Non-neutral Plasmas

While the previous section focused on an ideal Penning trap and single particle motion, the reality for ALPHA is more complicated. Adding many particles of the same charge into a single electrostatic well introduces collective effects that considerably alter motion. This section will deal with the motion of many particles in a Penning-Malmberg trap in a plasma state. Most plasmas in nature can be described as quasi neutral, namely they have a near balance of positively and negatively charged species. For the purposes of producing antihydrogen confinement of a non-neutral or charged plasma is required. This non-neutral plasma (NNP) lends itself to confinement in a Penning trap as we will shortly see and displays many of the same collective effects as a quasi neutral plasma.

2.2.1 *Defining a Plasma*

Next, it is important to determine a quantitative definition of a plasma and discuss some of the properties of a plasma. Being made up of charged particles that are able to reorder within the plasma, shielding of external fields occurs. One can also define the length scale over which electric fields are shielded by the plasma charges reordering. This length scale is known as the Debye length and must be much smaller than the radius of the cloud of charge for it to be considered a plasma [3]

$$\lambda_D \equiv \sqrt{\frac{\epsilon_0 T}{nq^2}}, \tag{2.8}$$

where λ_D is the Debye length, T is the temperature, n is the number density, q is the elementary charge, and ϵ_0 is the vacuum permittivity The frequency at which plasmas oscillate after reordering charges, known as the plasma frequency, is given by

$$\omega_p = \sqrt{\frac{nq^2}{\epsilon_0 m}}, \tag{2.9}$$

where n, q, ϵ_0 are as previously defined, and m is the mass of the particle. The plasma rotates in the magnetic field of the Penning trap as though rigid and has a uniform density, this density drops to zero on the scale of a Debye length.

2.2.2 *Confinement of Non-neutral Plasmas in a Penning-Malmberg Trap*

Confinement of a NNP in a Penning trap can be derived following [4], I will briefly summarise the essential points. As in the single particle example we set a magnetic field $\vec{B} = B_0\hat{z}$ and assume an electric potential $\phi(\vec{r})$ with $\vec{r} = (r, \theta, z)$,

$$\phi(\vec{r}) = \phi_T(\vec{r}) + \phi_P(\vec{r}). \tag{2.10}$$

The electric potential in the trap in the absences of the plasma, $\phi_T(\vec{r})$ satisfies Laplace's equation and is determined by the voltage applied to the electrodes. The overall potential has a term $\phi_P(\vec{r})$, accounting for the plasma self potential, or space charge, which is determined by the number of charges confined and the plasma density. This space charge can be large in ALPHA for electrons or positrons and can effectively reduce and flatten trapping potentials on the order of 10 V.

The motion of this collection of charges is then dictated by the following Hamiltonian in a cylindrical coordinate system,

$$H = \underbrace{\sum_{j=1}^{N}\left(\frac{p_{rj}^2}{2\,m} + \frac{[p_{\theta j} - \frac{q}{2}Br_j^2]^2}{2mr_j^2} + \frac{p_{zj}^2}{2\,m}\right)}_{\text{Kinetic energy}} + \overbrace{q\sum_{j=1}^{N}\phi_T(\vec{r_j})}^{\text{Interaction with trap potential}} + \underbrace{q\sum_{j=1}^{N}\phi_P(\vec{r_j})}_{\text{Interaction between charges}}, \tag{2.11}$$

with the source of the various contributions as labelled. The canonical momenta, $p_{rj}, p_{\theta j}, p_{zj}$ are defined as,

$$p_{rj} = m\frac{dr_j}{dt}, \qquad p_{\theta j} = mr_j^2\frac{d\theta_j}{dt} + \frac{qBr_j^2}{2}, \qquad p_{zj} = m\frac{dz_j}{dt}. \tag{2.12}$$

Given the cylindrical nature of the electrodes, there is a natural symmetry making the potential independent of θ (as long as ϕ_P is cylindrically symmetric), therefore $\phi(\vec{r}) \equiv \phi(r, z)$. A second consequence of this symmetry is that total canonical angular momentum, P_θ, is conserved.

$$P_\theta = \sum_{j=1}^{N} p_{\theta j} = \sum_{j=1}^{N} m r_j^2 \frac{d\theta_j}{dt} + \frac{qBr_j^2}{2} = L. \tag{2.13}$$

Since the Penning trap utilizes a strong magnetic field this equation reduces to

$$L = \frac{qB}{2} \sum_{j=1}^{N} r_j^2 = constant, \tag{2.14}$$

where L is the constant angular momentum, noting that this is only constant in a ideal Penning trap. As this equality means the mean squared radius of the plasma is conserved and radial confinement is achieved. If the plasma begins to radially expand and some particles are lost to the wall of electrode, most particles must remain at $r \approx 0$ to conserve the mean squared radius. This means only a small fraction (< 1 %) of particles can be lost radially if this initial mean square radius is sufficiently small.

As for axial confinement, it is comparatively trivial to see from the second term in Eq. 2.11 that provided the trapping potential, i.e. the voltage applied to the electrodes, is large then confinement can be achieved.

In a real Penning trap, imperfections in the electric or magnetic fields lead to expansion and heating of the plasma and must be combated through several plasma manipulation techniques to be discussed in Sects. 2.3.2 and 2.3.3.

2.3 Plasma Techniques in ALPHA

2.3.1 Plasma Temperature Diagnostic

Key parameters when making antihydrogen are the temperatures of the positron and antiproton plasmas. To determine the temperature of a plasma in ALPHA, particles are slowly released from a confining potential towards a microchannel plate detector (MCP) combined with a phosphor screen [5] and silicon photomultiplier (SIPM) [6]. We assume the plasma has a Maxwell distribution of energies of particles with mass m and velocity, v_i,

$$f(v_i) = \sqrt{\frac{m}{2\pi k_B T}} exp\left(-\frac{m v_i^2}{2k_B T}\right), \tag{2.15}$$

where k_B is the Boltzman constant and T is the temperature of the ensemble. The number of particles can that escape as the potential barrier height, V_B, is reduced is therefore proportional to,

$$N(V_B) \propto exp\left(-\frac{qV_B}{k_BT}\right). \tag{2.16}$$

By measuring the relative number of particles that escape as a function of barrier height the temperature can be extracted with a linear fit on a log scale. Positron and electron numbers are measured using a microchannel plate detector (MCP, as in Sect. 3.4.2), while antiproton numbers are measured by monitoring annihilations with a scintillator. There are several complicating factors with this technique related to three dimensional effects, such as hollowing of the plasma during release and heating as the well is changed ultimately limiting the number of charges that contain useful temperature information. For this reason the method here can only be successfully implemented for plasmas with large particle numbers (above 100k), alternatives must be sought for particle numbers less than 10k and will be discussed in Chap. 5.

2.3.2 *Plasma Cooling*

A major benefit of a Penning trap when confining electrons or positrons is cyclotron cooling [7]. As particles, with mass m and charge q, oscillate around the trap due to the magnetic field B, they emit photons and cool at a rate,

$$\frac{dT_\perp}{dt} = \frac{-T_\perp q^4 B^2}{3\pi \epsilon_0 m^3 c^3}, \tag{2.17}$$

where $T_\perp$ is the temperature transverse to the magnetic field, ϵ_0 is the vacuum permittivity and c is the speed of light.

Due to the inverse cubic dependence on mass, for more massive particles, such as antiprotons, this process is too slow so alternative cooling methods are required. In ALPHA, the preferred technique is sympathetic cooling using electrons or positrons to remove energy from the heavier particle through collisions, most commonly antiprotons cooled by electrons [8]. It is also possible to reduce the overall temperature of the plasma by removing more energetic particles. If one assumes the energy distribution of the particles in a plasma is a Maxwell-Boltzman distribution, removing the high energy tail of the distribution and allowing the plasma to thermalise results in a lower average energy, and therefore temperature, of the ensemble. This process is known as evaporative cooling and is carried out with manipulations of the electric fields to allow the most energetic particles to overcome the potential barrier [9].

2.3.3 *Reproducible Production of Plasmas with SDREVC*

It has been shown by ALPHA [10] and ATHENA [11] that production of antihydrogen varies significantly with different initial plasmas, as such obtaining reproducible antiproton and positron plasmas in a controlled fashion is essential. To control the radial size of a non-neutral plasma the rotating wall technique has been implemented in several Penning trap systems for many years [12]. The technique uses a segmented cylindrical electrode to apply an oscillating electric field to the plasma. The field couples to the plasma providing a torque and increases the rotational frequency when the electric field is rotating faster than the plasma rotation frequency, compressing the plasma [12]. An example of a six way segmented electrode (as used in ALPHA) can be seen in Fig. 2.1, this is a dipole mode where the applied potential is the same to each segment just offset by a phase of 60°. This compression can continue to a high plasma density where the rotation frequency is made to match the rotating wall drive frequency in what is known as the strong drive regime (SDR) [13].

In a development within the ALPHA collaboration, the rotating wall technique, in the strong drive regime, is combined with evaporative cooling (described in Sect. 2.3.2) of the plasma to fix both particle number and density [14]. The plasma is contained in a long (3–5 electrode) well with a segmented electrode at one end, the rotating wall is applied and the potential made more shallow. The space charge of the plasma is then set by the depth of the potential well at the end of this process, with one of the end confining potentials being smaller than the other as shown in Fig. 2.2 to control the direction of the evaporated particles. When the plasma is compressed in the strong drive regime it allows the production of a plasma with only 1% variability in particle number and plasma radius. Having reproducible plasmas over long periods of time has increased production of trappable antihydrogen and enabled detailed studies of magnetic fields [15] and patch potentials [16] using electron plasmas. In

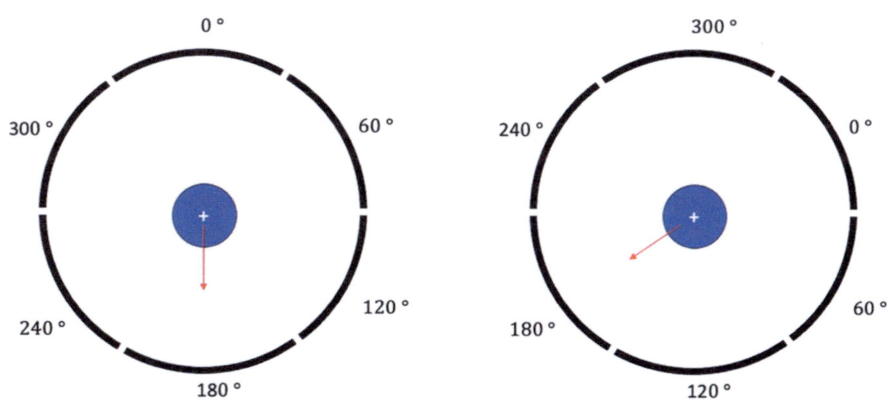

Fig. 2.1 A cartoon of 6-way segmented electrode with a positive non-neutral plasma (blue) being driven in the direction of the arrow (red) which oscillates at a set drive frequency

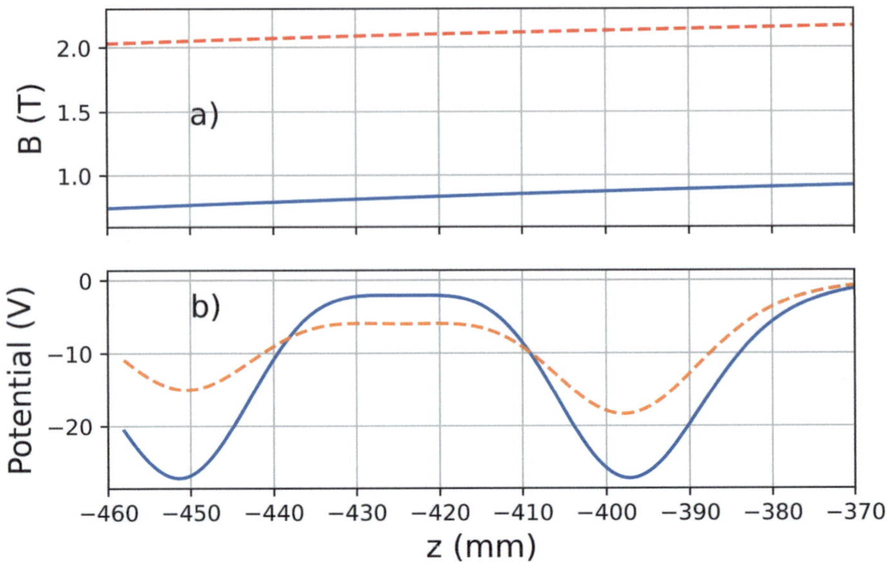

Fig. 2.2 Example electron reservoir SDREVC preparation used for ALPHA-g. **a** Magnetic field in the region, the blue line shows the nominal magnetic field in ALPHA-g with the red dashed line shows the boosted field from the capture solenoid (which will be discussed in Sect. 3.6) **b** Shows the electrostatic potentials used at the start of the rotating wall (blue) and end of the EVC process (orange dashed)

a development for the ALPHA-g experiment, this strong drive regime evaporative cooling (SDREVC) technique was demonstrated in a significantly lower magnetic field, this will be discussed further in Chap. 5.

2.4 Antihydrogen Formation

The first observations of antihydrogen were in particle accelerator experiments as mentioned previously with LEAR [17] and Fermilab [18]. These antihydrogen atoms were formed in a small fractions of collisions of protons with heavy atomic nuclei and were highly energetic so had a short lifetime before annihilating. For the purpose of the study of antihydrogen, low energy atoms that can be studied over a long period of time are required.

In the previous section we have shown how to produce cold positron and antiproton plasmas with desired size and density in a highly reproducible way. To form antihydrogen, antiproton and positron plasmas are slowly merged by manipulations of their respective electrostatic confining potentials as shown in Fig. 2.3.

Several mechanisms for antihydrogen production in ALPHA have been suggested, the most likely of these is three-body recombination as shown in Eq. 2.18. In this

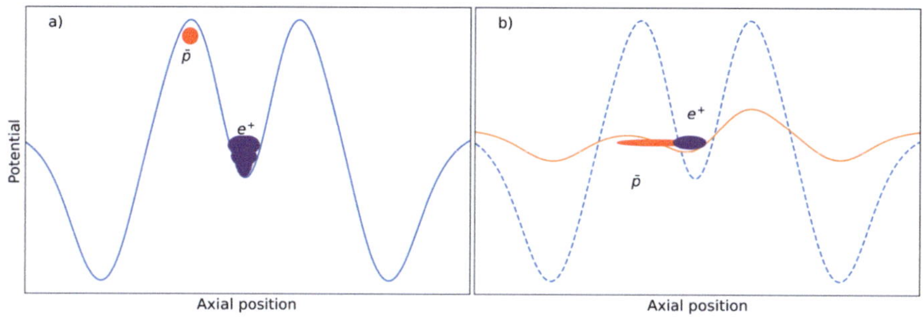

Fig. 2.3 An illustration of the slow mixing process used to form antihydrogen in ALPHA. **a** Electrostatic confining potentials for antiprotons (red) and positrons (purple) before mixing (blue), **b** Electrostatic potentials during mixing (orange) where antiprotons are injected into the positron plasma

process a pair of positrons collide in the vicinity of an antiproton, the collision allows one positron to become low enough energy to form the bound state of antihydrogen while the remaining positron carries off the excess energy

$$e^+ + e^+ + \bar{p} \rightarrow \bar{H} + e^+. \tag{2.18}$$

Radiative recombination is also believed to play a role and is the inverse process to photoionisation, relying on the spontaneous emission of a photon to carry energy and momentum away, allowing antihydrogen formation as follows,

$$\bar{p} + e^+ \rightarrow \bar{H} + \gamma. \tag{2.19}$$

While these are the accepted mechanism for formation in ALPHA, there are several antihydrogen formation mechanisms relied upon by the experiments discussed in Chap. 1. Charge exchange requires forming positronium and relying on the electron to be replaced by incoming antiprotons and remove energy as in,

$$\bar{p} + Ps \rightarrow \bar{H} + e^-. \tag{2.20}$$

This has been experimentally shown by ATRAP [19] and more recently in 2023 by the GBAR experiment [20]. Using an excited state positronium during a charge exchange formation, as in

$$\bar{p} + Ps^* \rightarrow \bar{H}^* + e^-, \tag{2.21}$$

increases the cross section as it has a longer lifetime due to smaller overlap of positron and electron and the AEGIS experiment has show successful production of antihydrogen through this technique [21].

2.5 Antihydrogen Trapping

While the Penning trap is a powerful tool, antihydrogen being neutral requires a different approach. A magnetic minimum trap, known as an Ioffe-Pitchard trap or magnetic bottle, can be used to confine any particle or atom with a magnetic moment [22]. Antihydrogen is a diamagnetic atom with a magnetic moment $\vec{\mu}$ and has states that are trappable in the magnetic minimum in a potential,

$$U = -\vec{\mu} \cdot \vec{B}. \tag{2.22}$$

with

$$\vec{\mu} \simeq -2\pi g_j \mu_B \frac{\vec{J}}{h}, \tag{2.23}$$

g_j is the Lande g-factor, $\vec{J}$ is the total angular momentum of the positron, μ_B is the Bohr magneton and h is Plank's constant. This creates a confining force

$$\vec{F} = \nabla(\vec{\mu} \cdot \vec{B}), \tag{2.24}$$

which has a spin dependence. In a high magnetic field, antihydrogen with the positron spin aligned with the magnetic field are in an un-trappable high field seeking state. For spin anti-aligned with the magnetic field, the antihydrogen is in a low field seeking state and is trappable in a three dimensional minima formed of end solenoids for axial confinement and a multi-pole magnet, such as a quadrupole, to produce the required radial field gradient.

The energy level structure for the 1S manifold of hydrogen can be seen in Fig. 2.4. The previously described formation of antihydrogen in ALPHA leads to a small fraction of atoms that are of low enough energy to be confined. The threshold energy, also to be thought of as the trap depth is calculated to be,

$$\Delta U = 0.67[\mathrm{K}]\Delta B \tag{2.25}$$

where ΔB is the difference between the magnetic field minima and maxima in Tesla. The units are given in Kelvin as this is a convenient energy scale for trapped low energy atoms.

As the magnetic fields from the neutral atom trap break the magnetic symmetry of the Penning trap, leading to expansion and heating of plasmas, ALPHA uses an octupole for radial confinement rather than a quadrupole to minimise this effect. The use of an octupole however means it is difficult to achieve a radial field gradient greater than 1 T as the radial field strength decrease as $1/r^3$.

In ALPHA-g, 3×10^6 positrons are slowly merged with approximately 5×10^4 antiprotons to form a few thousand antihydrogen, on average two of these are trapped.

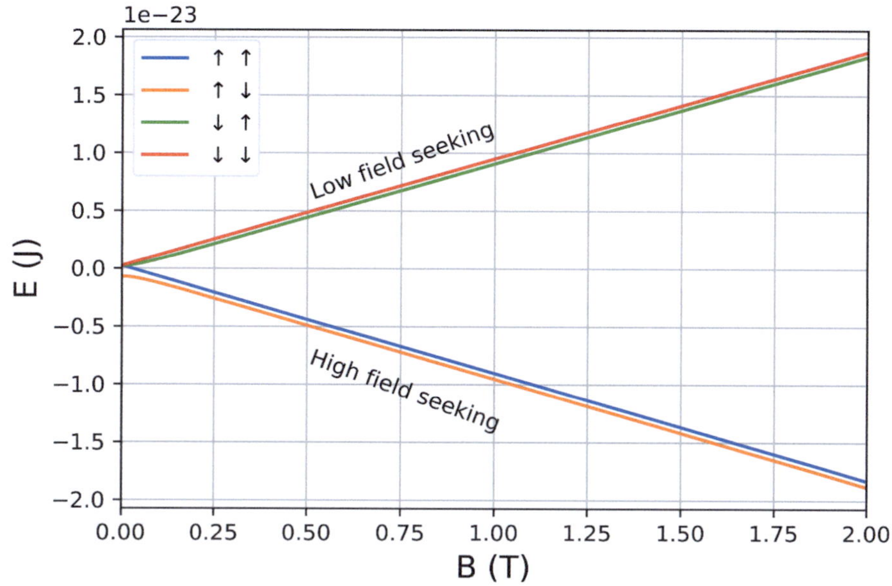

Fig. 2.4 States of the 1S manifold in (anti)hydrogen. The first arrow in the legend shows electron (positron) spin, the second (anti)proton spin

Given the small number of antihydrogen trapped per mixing, it is beneficial to be able to accumulate antihydrogen over many of these mixing events and perform experiments with a large signal fewer times, reducing background contaminants. The process of accumulation or "stacking" of antihydrogen was established by ALPHA [23] and has allowed a step change in spectroscopy precision as well as the gravity measurement shown in Chap. 6. Keeping all the superconducting magnets required for trapping energized for up to 24 h allows for a more stable magnetic environment and the larger sample yields a result in a single day that previous required many weeks [24]. Efforts are underway to increase the number of antihydrogen produced per mixing, one example is through sympathetic cooling of positron plasmas using laser cooled Beryllium ions [25].

This chapter has provided the background necessary to understand the methodology behind the ALPHA-g free fall measurement that will be shown in Chap. 6. The next chapter aims to give the reader an overview of the ALPHA apparatus so one can follow the journey of positrons and antiprotons as they form antihydrogen and are trapped then studied.

References

1. Hodgkinson DL (2022) On the dynamics of adiabatically cooled antihydrogen in an octupole-based Ioffe-Pritchard magnetic trap. PhD thesis. The University of Manchester (United Kingdom)
2. Friesen TP (2014) Probing trapped antihydrogen: in situ diagnostics and observations of quantum transitions. [Online; accessed 17. Sep. 2024]. PhD thesis. University of Calgary. https://alpha.web.cern.ch/sites/default/files/2020-05/FriesenThesis.pdf
3. Meyer-Vernet N (1993) Aspects of Debye shielding. Am J Phys 61(3):249–257. ISSN: 0002-9505. https://doi.org/10.1119/1.17300. eprint: https://pubs.aip.org/aapt/ajp/article-pdf/61/3/249/12179178/249/_1/_online.pdf
4. Dubin DHE, O'Neil TM (1999) Trapped nonneutral plasmas, liquids, and crystals (the thermal equilibrium states). Rev Mod Phys 71(1):87–172. https://doi.org/10.1103/RevModPhys.71.87. https://link.aps.org/doi/10.1103/RevModPhys.71.87
5. Andresen GB et al (2009) Antiproton, positron, and electron imaging with a microchannel plate/phosphor detector. Rev Sci Instrum 80(12):123701. ISSN: 0034-6748. https://doi.org/10.1063/1.3266967. eprint: https://pubs.aip.org/aip/rsi/article-pdf/doi/10.1063/1.3266967/16070067/123701/_1/_online.pdf
6. Hunter ED et al (2020) Plasma temperature measurement with a silicon photomultiplier (SiPM). Rev Sci Instrum 91(10):103502. ISSN: 0034-6748. https://doi.org/10.1063/5.0006672. eprint: https://pubs.aip.org/aip/rsi/article-pdf/doi/10.1063/5.0006672/14878456/103502/_1/_online.pdf
7. O'Neil TM (1980) Cooling of a pure electron plasma by cyclotron radiation. Phys Fluids 23(4):725–731. ISSN: 0031-9171. https://doi.org/10.1063/1.863044. eprint: https://pubs.aip.org/aip/pfl/article-pdf/23/4/725/12267881/725/_1/_online.pdf
8. Andresen GB et al (2008) Compression of antiproton clouds for antihydrogen trapping. Phys Rev Lett 100(20):203401. https://doi.org/10.1103/PhysRevLett.100.203401. https://link.aps.org/doi/10.1103/PhysRevLett.100.203401
9. Andresen GB et al (2010) Evaporative cooling of antiprotons to cryogenic temperatures. Phys Rev Lett 105(1):013003. https://doi.org/10.1103/PhysRevLett.105.013003. https://link.aps.org/doi/10.1103/PhysRevLett.105.013003
10. Andresen GB et al (2010) Trapped antihydrogen. Nature 468:673–676. ISSN: 1476-4687. https://doi.org/10.1038/nature09610
11. Amoretti M et al (2002) Production and detection of cold antihydrogen atoms. Nature 419:456–459. ISSN: 1476-4687. https://doi.org/10.1038/nature01096
12. Huang X-P et al (1997) Steady-state confinement of non-neutral plasmas by rotating electric fields. Phys Rev Lett 78(5):875–878. https://doi.org/10.1103/PhysRevLett.78.875. https://link.aps.org/doi/10.1103/PhysRevLett.78.875
13. Danielson JR, Surko CM (2006) Radial compression and torque-balanced steady states of single-component plasmas in Penning-Malmberg trapsa). Phys Plasmas 13.5 (May), p. 055706. ISSN: 1070-664X. https://doi.org/10.1063/1.2179410. eprint: https://pubs.aip.org/aip/pop/article-pdf/doi/10.1063/1.2179410/15798675/055706/_1/_online.pdf
14. Ahmadi M et al (2018) Enhanced control and reproducibility of non-neutral plasmas. Phys Rev Lett 120(2):025001. https://doi.org/10.1103/PhysRevLett.120.025001. https://link.aps.org/doi/10.1103/PhysRevLett.120.025001
15. Hunter ED et al (2020) Electron cyclotron resonance (ECR) magnetometry with a plasma reservoir. Phys Plasmas 27(3):032106. ISSN: 1070-664X. https://doi.org/10.1063/1.5141999. eprint: https://pubs.aip.org/aip/pop/article-pdf/doi/10.1063/1.5141999/13806318/032106/_1/_online.pdf
16. Baker CJ et al (2024) Measurements of Penning-Malmberg trap patch potentials and associated performance degradation. Phys Rev Res 6(1):L012008. https://doi.org/10.1103/PhysRevResearch.6.L012008. https://link.aps.org/doi/10.1103/PhysRevResearch.6.L012008

17. Baur G et al (1996) Production of antihydrogen. Phys Lett B 368(3):251–258. ISSN: 0370-2693. https://doi.org/10.1016/0370-2693(96)00005-6. https://www.sciencedirect.com/science/article/pii/0370269396000056
18. Blanford G et al (1998) Observation of atomic antihydrogen. Phys Rev Lett 80(14):3037–3040. 10.1103/PhysRevLett.80.3037. https://link.aps.org/doi/10.1103/PhysRevLett.80.3037
19. Fitzakerley DW (2015) Antihydrogen via two-stage charge exchange. [Online; accessed 16. Sep. 2024]. PhD thesis. York University. https://yorkspace.library.yorku.ca/items/c3ac3700-8c37-4fb2-91ff-f40ac956f97b/full
20. Adrich P et al (2023) Production of antihydrogen atoms by 6 keV antiprotons through a positronium cloud. arXiv:2306.15801 [hep-ex]
21. Amsler C et al (2021) Pulsed production of antihydrogen. Commun Phys 4(19):1–11. ISSN: 2399-3650. https://doi.org/10.1038/s42005-020-00494-z
22. Pritchard DE (1983) Cooling neutral atoms in a magnetic trap for precision spectroscopy. Phys Rev Lett 51(15):1336–1339. https://doi.org/10.1103/PhysRevLett.51.1336. https://link.aps.org/doi/10.1103/PhysRevLett.51.1336
23. Ahmadi M et al (2017) Antihydrogen accumulation for fundamental symmetry tests. Nat Commun 8(681):1–6. ISSN: 2041-1723. https://doi.org/10.1038/s41467-017-00760-9
24. Baker CJ et al. (2021) Laser cooling of antihydrogen atoms. Nature 592:35–42. ISSN: 1476-4687. https://doi.org/10.1038/s41586-021-03289-6
25. Baker CJ et al (2021) Sympathetic cooling of positrons to cryogenic temperatures for antihydrogen production. Nat Commun 12(6139):1–8. ISSN: 2041-1723. https://doi.org/10.1038/s41467-021-26086-1

Chapter 3
Apparatus

The theoretical foundations for the ALPHA Penning-Malmberg and magnetic minimum traps have been laid out in the previous chapter. The combination of these two traps allows the confinement of charged particles, the manipulation of these charged particles to form antihydrogen and finally the confinement of neutral antihydrogen.

This chapter will outline the sections of the ALPHA apparatus necessary to describe gravity experiments in ALPHA-g. ALPHA is a well established experiment in the AD facility and has provided many breakthroughs in the field including the first trapping and spectroscopy of antihydrogen [1, 2]. In a rapid expansion since 2018 to include a new experiment, ALPHA-g, the collaboration now aims to observe the free fall of antihydrogen in Earth's gravitational field. The technical challenges related to ALPHA-g will become clear throughout this chapter as the issues of homogeneous magnetic fields, working with an ultra-high vacuum system and cryogenics and many more issues that are complicated further by the scale of the apparatus. The whole apparatus can be seen in Fig. 3.1, on either end are the preparation areas for either antiprotons in the catching trap or positrons in the positron accumulator. The two regions for carrying out experiments can be seen as either ALPHA-2, the apparatus used for spectroscopy [3], or ALPHA-g, the previously mentioned gravity experiment.

3.1 Catching Trap

Antiprotons are delivered to ALPHA from ELENA through an electrostatic beamline at 100 keV, with a bunch of 9 million arriving every 4 min. Once the antiprotons reach ALPHA, the bunch is steered through a pure aluminum foil to further reduce the beam energy [4], antiprotons that do not reach the foil annihilate on vacuum chamber walls. Antiprotons with an energy below 5 keV can be trapped in the first of many ALPHA Penning-Malmberg traps known as the "Catching trap", seen on the far

A. M. W. Powell, *Magnetic Field Characterisation for Gravitational Free Fall Measurements of Antihydrogen in the ALPHA-g Experiment*, Springer Theses,
https://doi.org/10.1007/978-3-032-07190-3_3

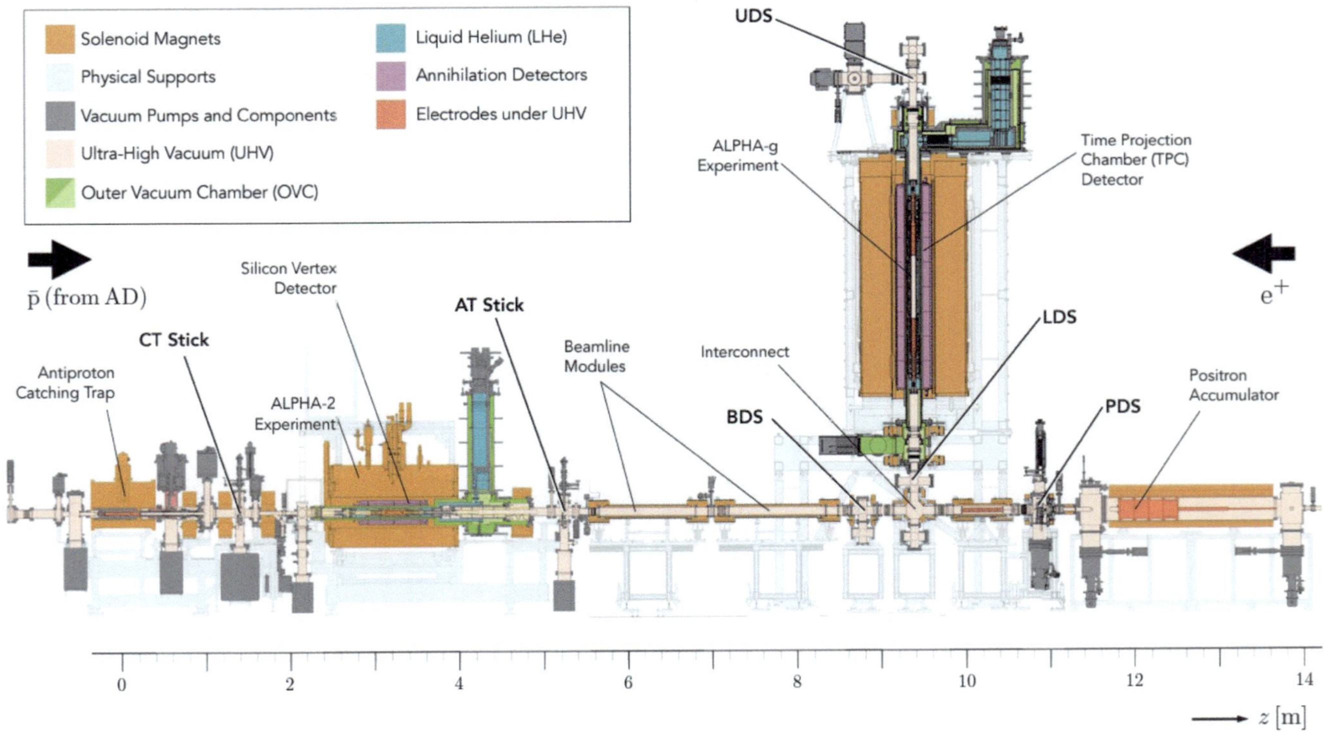

Fig. 3.1 A schematic of the whole ALPHA apparatus with the connection to ELENA (see Fig. 1.2) on the left and the positron source on the right. Particle diagnostic stations as described in Sect. 3.4 are shown as Catching trap stick (CT stick), atom trap stick (AT stick), beamline diagnostic station (BDS), Positron diagnostic station (PDS), lower diagnostic station (LDS) and upper diagnostic station (UDS)

left of Fig. 3.1. This trap utilizes fast switching high voltage electrodes, synchronized to the arrival of the antiproton pulse, in a 3 T magnetic field to confine the antiprotons after they have entered the trap, effectively closing an electrical potential gate behind the bunch. The antiproton bunch is caught in an electrostatic well with electrons pre-loaded to allow sympathetic cooling, down to approximately 100 K, in preparation for distribution to other areas of the experiment. The entire catching trap assembly was modified in 2021 to be compatible with the new lower energy antiproton source from ELENA, the details of this upgrade are discussed elsewhere [4]. Typically the antiproton bunch after catching trap preparation in 2022 consisted of around 5×10^5 antiprotons, though, due to calibration challenges with the detectors around this trap, this is only an estimate.

3.2 Positron Accumulator

The positrons used to form antihydrogen in ALPHA are produced in a radioactive sodium-22 source via beta plus decay,

$$^{22}_{11}Na \rightarrow {}^{22}_{10}Ne + e^{+} + \nu_{e}. \tag{3.1}$$

The source has a solid neon moderator with an applied bias and is cooled to 5–8 K. The moderator is used to provide a positron beam with a narrower energy distribution than is naturally occurring through the decay [5]. Positrons at the moderated energy are magnetically steered into a three stage Surko style buffer gas positron accumulator [7] on the far right of Fig. 3.1. The accumulation occurs in a Penning-Malmberg trap with variable electrode diameters allowing nitrogen buffer gas held at variable pressure across the accumulation trap as shown in Fig. 3.2. Positrons cool through collisions with the buffer gas until they are captured in the lowest pressure region due to blocking electrostatic potentials on the Penning trap electrodes reducing the axial confinement volume as the positrons cool.

Typically in 2022, ALPHA accumulated positrons for 100 s and had approximately 14×10^6 available for transfer to ALPHA-g. As the buffer gas trap is not cryogenic the positron bunch is typically at room temperature before transfer. This number of

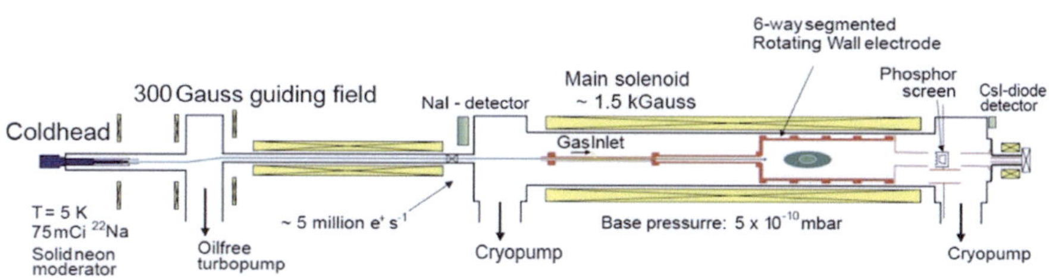

Fig. 3.2 A schematic of the Surko buffer gas positron accumulator in ALPHA [6]

particles is set by the accumulation efficiency, which is typically 2–5% in a buffer gas trap. The accumulation time is largely determined by the activity of the sodium-22 radionuclides, which have a half life of ≈ 2.5 years and the source had been in place for several years at this point.

3.3 Beamline

Now that we have sources of both positrons and antiprotons at low energy, it is time to address transporting particles between Penning traps in ALPHA, a major challenge in the expanded ALPHA landscape. Moving low energy particles through multiple large field gradients often leads to losses due to magnetic mirroring or radial expansion. In 2018, a new beamline was added to ALPHA to accommodate the ALPHA-g apparatus. This beamline consists of seven sections of normal-conducting, water-cooled, magnets with each section having multiple solenoids and can be seen in Fig. 3.1. The solenoids in the beamline are multi-directional to allow for the transfer of positrons and antiprotons to ALPHA-2 or ALPHA-g. The most complex section of the beamline is known as the interconnect and steers antiprotons and positrons around the ninety degree turn into ALPHA-g as seen in Fig. 3.3. The simulation, design, and performance of this new beamline are summarised in [8].

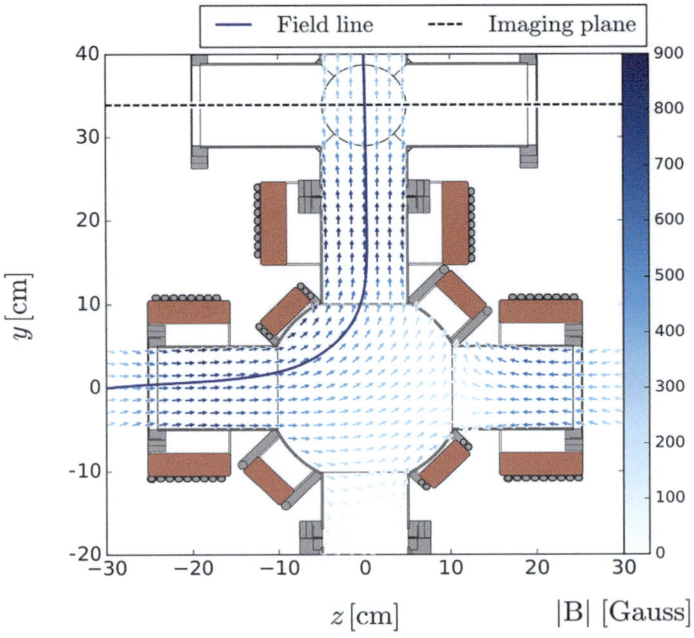

Fig. 3.3 Example of magnetic field lines experienced by particles being transferred from ALPHA-2 to ALPHA-g in the interconnect region of the beamline [8]

3.4 Vacuum Manipulator Particle Diagnostic Stations

To aid in the transport of particles, diagnostic stations have been assembled at different points along the experiment as seen in Fig. 3.1. The diagnostic stations rely on being able to move different instruments into the beam path in the vacuum chamber to monitor the beam properties. Using a vacuum bellows connected to a stepper-motor instruments can be moved into position in the center of the beamline. An example diagnostic station known as the positron diagnostic station (PDS) can be seen in Fig. 3.4 and a summary of instruments on each of the diagnostic stations can be found in Table 3.1.

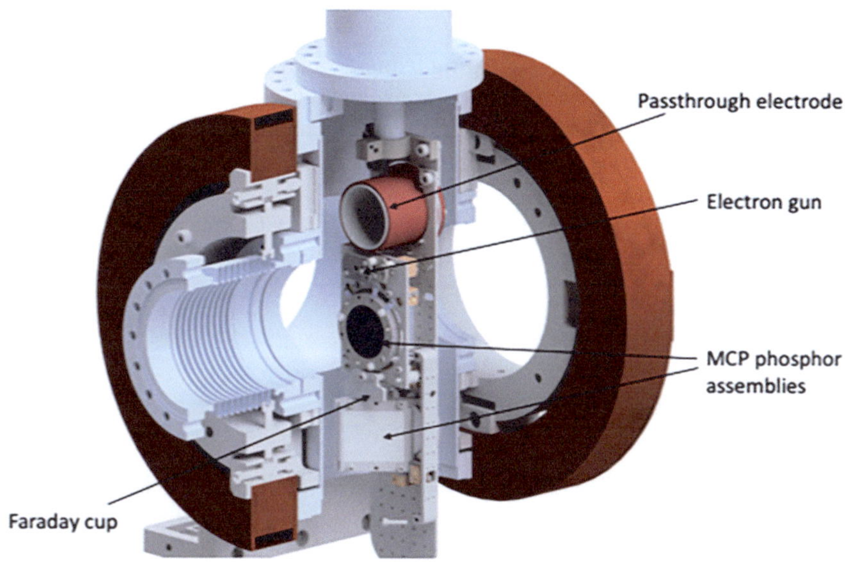

Fig. 3.4 Example diagnostic station with vertical vacuum manipulator and various instruments. Adapted with permission from [8]

Table 3.1 Summary of diagnostic station instrumentation

Instrument	USCT	DSCT	USAT	DSAT	BDS	LDS	UDS	PDS	PB2
Egun	×	✓	✓	✓	✓	✓	✓	✓	×
FC	×	✓	×	✓	✓	✓	×	✓	×
MCP	×	✓	✓	✓	✓	✓	✓	✓	×
Phosphor	✓	×	×	×	×	×	×	×	✓
Special	–	–	Be target	–	–	–	Horn	–	–

3.4.1 *Electron Source and Electron Loading*

A standard barium oxide filament is used to produce thermal electrons when a current is applied. These electrons are collimated and accelerated by a voltage difference between two plates to produce a stable beam of electrons at a set energy, this source is known as an e-gun. Through a feedback loop, the emission current is regulated by making small adjustments to the current applied in response to fluctuations, allowing for consistent electron beam over time.

3.4.2 *Microchannel Plates and Phosphor Screens*

A key diagnostic tool is particle imaging using a microchannel plate (MCP) combined with a phosphor screen and detector as depicted in the cartoon Fig. 3.5. The charged particles are accelerated through the microchannels and produce a cascade of electrons, these electrons are accelerated to the phosphor screen, which scintillates and the light is reflected by a 45° mirror and detected by a charge-coupled device (CCD) or silicon photomultiplier (SIPM) [9, 10]. The CCD allows for 2d imaging of charged particle clouds and the SiPM allows for sensitive measurements of the relative number of particles reaching the MCP (Sect. 2.3.1)

3.4.3 *Faraday Cup*

To estimate the number of particles at a given stage of the plasma preparation it is useful to be able to measure the charge. In a Faraday cup (FC) diagnostic, charged particles are ejected from the trap to a metal cup on the diagnostic station. When the

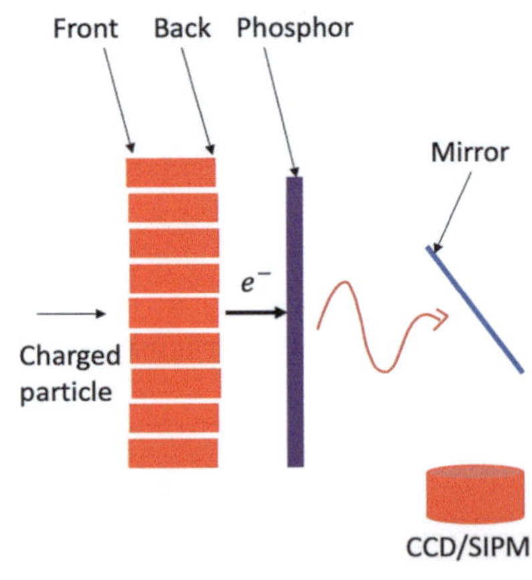

Fig. 3.5 Cartoon stack up of an MCP and phosphor screen assembly in ALPHA used for charged particle detection

charged particles hit the cup they induce a small charge, which can be measured and the number of charges present is calculated. For practical purposes, the function of the FC is often replaced by the front of the MCP. When no voltage is applied the front plate of the MCP can measure the same charge but with the added benefit of being much larger and therefore much easier to guarantee that all of the particles are accounted for in the measurement.

3.5 ALPHA-g

The basic concept for ALPHA-g, which will be discussed in Chap. 6, is to make a vertical antihydrogen trap with a significant gravitational potential difference at the ends of the magnetic trap. Practical measures to improve the magnetic environment for a free fall experiment have led to complex magnet systems but a cartoon layout of the apparatus can be seen in Fig. 3.6 with details of each section to be shown.

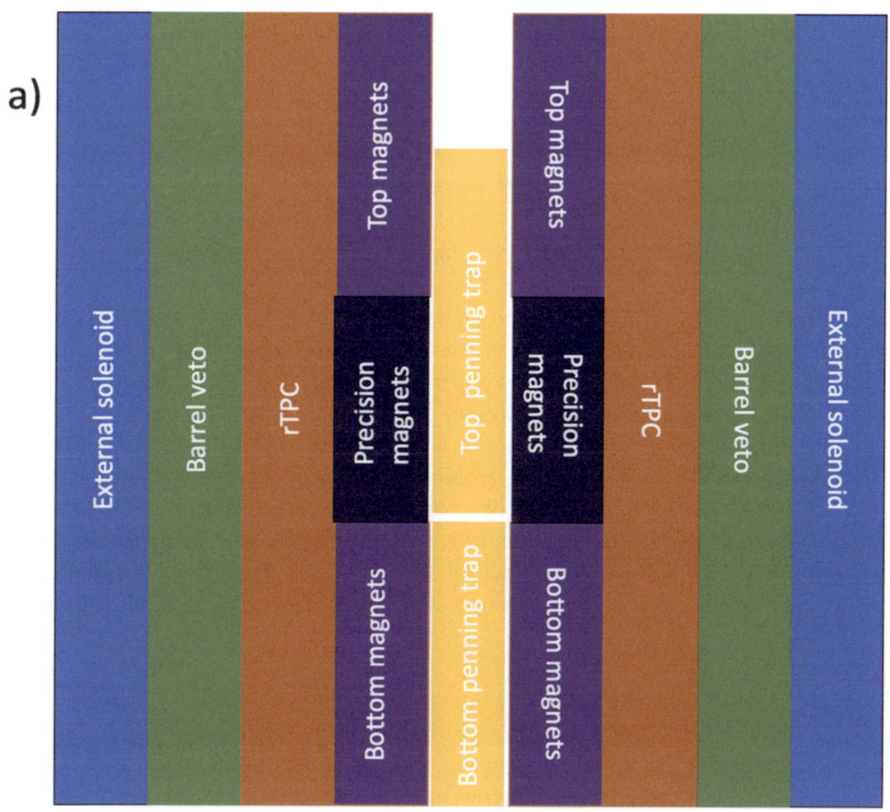

Fig. 3.6 Cartoon diagram showing a cut through of the cylindrical layers of ALPHA-g

Note that the top Penning trap is not yet installed but has been designed as will be discussed in Sect. 4.1.4.

3.5.1 Babcock Magnet

An external superconducting solenoid produced by Babcock, provides a homogeneous magnetic field for the ALPHA-g Penning trap at around 1 T. At 3 m length with a 50 cm diameter bore, it is designed to provide a large homogeneous region in the middle of the magnet. The magnet is cooled to a superconducting state by pulse tube cryocoolers with a closed helium system and compressors. The cooling power offered by this equipment has been marginal for the size of the magnet and has led to significant issues for ALPHA-g which are discussed elsewhere [11]. Improvements to the cooling system are underway in 2023 ahead of future ALPHA-g experiments. Operationally, this magnet is run connected to its power supply with active quench protection. A separate power supply is used for shimming coils, added in 2021 for improved homogeneity based on field mapping with Hall/NMR probes [12]. Extensive magnetic field characterisation in the Penning trap region was carried out and will be shown in Chap. 5.

3.5.2 Radial Time Projection Chamber (rTPC)

The rTPC is the main detector used in reconstructing the annihilations of antiprotons and antihydrogen atoms in ALPHA-g. It is a gas-filled radial time projection chamber using a mixture of carbon dioxide and argon continuously flowing around the anode wires. Charged particles ionize the gas, leaving free electrons, which are accelerated to the anode wires. The current flowing in the anode wires then induce an image charge on the cathode pads. The timing of these image charges on the cathode pads allow for reconstruction of the track of the annihilation products through the detector. Since there are multiple charged pions produced during an antiproton annihilation, the combined tracks can be used to reconstruct an annihilation position [13]. This position reconstruction is key to the magnetic release experiments, which will be discussed in Chap. 6. Figure 3.7 shows the trajectory of a pion through the gas detector (green) and the path of electrons drifting through the electric and magnetic fields (orange).

3.5.3 Barrel Veto

As with all particle detection experiments, background rejection is vital, with the main source of background in ALPHA being cosmic rays. The barrel veto is a detector positioned around the rTPC made with 64 trapezoidal scintillator bars with arrays

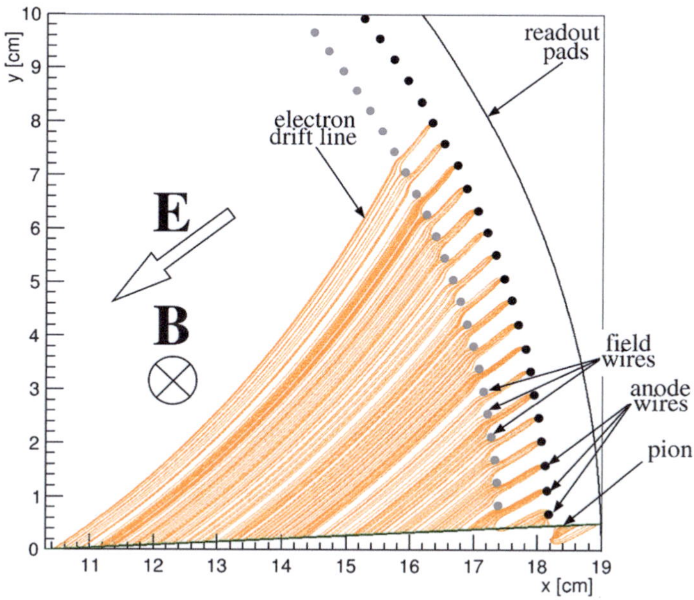

Fig. 3.7 Example simulated track of a charged pion (green) in the rTPC, showing trajectories of electrons produced (orange) [13]

of SiPM at each end [14]. The barrel uses coincidence triggers on multiple bars to distinguish a cosmic signal from an annihilation event (a cosmic ray would only trigger two bars as it passes through the detector in a straight line). This detector can also use the time of triggers on different bars to further increase the rejection capabilities, the timing of opposing hits on the scintillating bars can be used to distinguish external cosmic ray passages from pions originating in ALPHA-g, this is a complicated process due to the fast read-out requirements. Work is in progress to add this additional time of flight capability. The barrel veto was used in the reconstruction of annihilation events for data shown in Chap. 6 using coincidence events across multiple bars as an indicator of antihydrogen annihilation.

3.6 ALPHA-g Magnet System

3.6.1 Top and Bottom Magnetic Traps

The magnet system in ALPHA-g is conceptually the same as previous ALPHA devices with a magnetic trap formed of octupoles and short solenoids that are wound directly onto a steel bore that is connected to the cryostat and sealed, allowing the magnets to be submerged in liquid helium. However, for ALPHA-g minimising

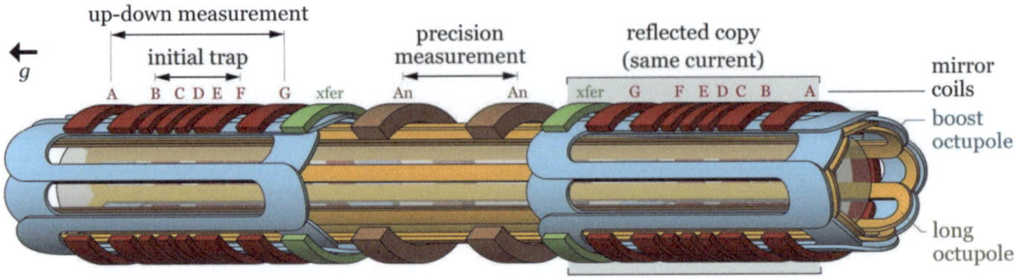

Fig. 3.8 Schematic of the ALPHA-g magnets forming three magnetic minima traps, image credit Dr. C.So

uncontrolled magnetic fields is critical to any successful measurements, it is particularly problematic to have asymmetric uncontrolled fields along the vertical axis. These uncontrolled fields arise due to magnetic field changes in the presence of superconducting material and will be discussed in Chap. 5. To reduce these issues, ALPHA-g has three magnetic traps, a top trap, a bottom trap, and a precision trap as in Fig. 3.8.

The top and bottom traps are mirror images of one another centered around the precision trap and each consist of a short octupole, and seven mirror coils (labeled A through G) [15]. Along with the magnets used in the magnetic minimum trap itself, each trap has a capture solenoid to boost the field by 2 T for particle preparation and transfer coils to boost low-field regions to improve particle transfer between magnetic traps. The intention here is to have symmetric placement of any superconducting material and the mirrored traps allow for systematic checks by performing the same experiment in two nominally identical systems at different heights. This also allows for identical magnet ramps to occur in the end traps before any work in the precision region takes place, making any uncontrolled fields symmetric.

3.6.2 Precision Magnetic Trap

The so-called precision region as shown in the center of Fig. 3.8 consists of analysis coils, the long octupole and some corrector coils not shown. It removes as much superconducting material as possible again reducing the size of any uncontrolled field and is discussed in detail in [15]. This region can only confine lower-energy antihydrogen atoms, as the total field gradient is smaller, which are more sensitive to the gravitational effects when released. This means production of the trapped sample is best made in the top or bottom trap and transferred into the precision region, allowing for adiabatic cooling during transfer [16]. The removal of a radial step in the bottom Penning trap and the implications for antihydrogen transfer will be discussed in Sect. 4.1.2. It is worth noting the implementation of laser cooling of antihydrogen, previously demonstrated in the ALPHA-2 apparatus, in ALPHA-g will also allow for an increased yield of this transfer.

The first design of a Penning trap to occupy this precision region and allow for magnetic field characterisation will be discussed in Sect. 4.1.4 and the prospect of reaching a high precision measurement using this set of magnets will be discussed in Sect. 8.2.

3.6.3 Magnet System Relevant for this Thesis

In the bottom magnetic trap region, a Penning trap has been inserted for antihydrogen production. The whole bottom magnetic and Penning trap can be seen in Fig. 3.9. Particular attention was paid to ensuring an electrode was centered under each of the end solenoids and the central solenoid for magnetometry purposes, as will be discussed in Chaps. 5 and 7. A summary of the magnets used in ALPHA-g for the data shown in this thesis can be seen in Table 3.2 with their power supplies and levels of control/noise. The mirror coils were connected in series and run with the same power supply to minimise current noise difference between the coils. To apply a magnetic bias, as will be explained in Sect. 6.1, a second power supply was connected to the top mirror coil (Mirror G). This power supply is low noise (0.4 mA) and operated at low currents (max 3 A).

During the 2022 campaign mirror coils B–F were not used, however they were commissioned and their current regulated at 0 A. There are plans to have varied field shapes during stacking in the future to again allow for adiabatic cooling or spectroscopy but these were not necessary for the first studies in ALPHA-g.

This chapter has summaries the key hardware components of the ALPHA apparatus used to prepare positrons and antiprotons for antihydrogen formation. It has then discussed elements of ALPHA-g used to form the magnetic minimum trap and used for annihilation detection. There are two essential hardware items missing from

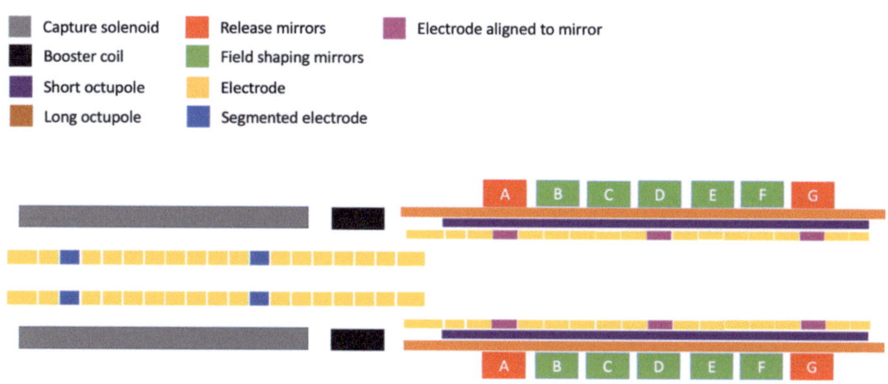

Fig. 3.9 Cartoon of the location of magnets and electrodes in the bottom section of ALPHA-g

Table 3.2 Summary of magnets forming the magnetic trap used in release experiments shown in Chap. 6

Magnet	Supply name	Operating current (A)	Current programming resolution (mA)	Current noise (mA rms)
Long octupole (LOc)	Sorensen SGA 10-1200	830	37	0.5
Bottom octupole (OcB)	Sorensen SGA 10-1200	830	37	0.6
Mirrors (MAG)	CAENELS FAST-PS-1K5	70	3.1	0.7
Mirror G Bias (MGDiff)	Kepco BOP 20-10	3	0.34	0.4
External solenoid main coil	2 x CAENELS FAST-PS-1K5	191	0.8	1.8
External solenoid shim coils	CAENELS FAST-PS-1020-200	5	0.1	1.5

the discussion so far, the ALPHA-g Penning trap electrodes and microwave system. These systems have undergone significant design upgrades from former versions within ALPHA which will be shown in the next chapter.

References

1. Andresen GB et al (2010) Trapped antihydrogen. Nature 468:673–676. ISSN: 1476-4687. https://doi.org/10.1038/nature09610
2. Amole C et al (2012) Resonant quantum transitions in trapped antihydrogen atoms. Nature 483:439–443. ISSN: 1476-4687. https://doi.org/10.1038/nature10942
3. Ahmadi M et al (2018) Characterization of the 1S–2S transition in antihydrogen. Nature 557:71–75. ISSN: 1476-4687. https://doi.org/10.1038/s41586-018-0017-2
4. Fabbri S (2023) Optimization of antiproton capture and delivery for the ALPHA antihydrogen experiment. [Online; accessed 26. Aug. 2023]. PhD thesis. The University of Manchester. https://research.manchester.ac.uk/en/studentTheses/optimization-of-antiproton-capture-and-delivery-for-the-alpha-ant
5. Gullikson EM, Mills AP, McRae EG (1988) Observation of intense two-beam positron diffraction and the precise determination of the positron band gap in rare-gas crystals. Phys Rev B 37(1):588–591. https://doi.org/10.1103/PhysRevB.37.588. https://link.aps.org/doi/10.1103/PhysRevB.37.588
6. ALPHA collaboration. ALPHA webpage. https://alpha.web.cern.ch/component/positron-accumulator
7. Murphy TJ, Surko CM (1992) Positron trapping in an electrostatic well by inelastic collisions with nitrogen molecules. Phys Rev A 46(9):5696–5705. https://doi.org/10.1103/PhysRevA.46.5696. https://link.aps.org/doi/10.1103/PhysRevA.46.5696

8. Baker CJ et al (2023) Design and performance of a novel low energy multispecies beamline for an antihydrogen experiment. Phys Rev Accel Beams 26(4):040101. https://doi.org/10.1103/PhysRevAccelBeams.26.040101. https://link.aps.org/doi/10.1103/PhysRevAccelBeams.26.040101
9. Andresen GB et al (2009) Antiproton, positron, and electron imaging with a microchannel plate/phosphor detector. Rev Sci Instrum 80(12):123701. ISSN: 0034-6748. https://doi.org/10.1063/1.3266967. eprint: https://pubs.aip.org/aip/rsi/article-pdf/doi/10.1063/1.3266967/16070067/123701/_1/_online.pdf
10. Hunter ED et al (2020) Plasma temperature measurement with a silicon photomultiplier (SiPM). Rev Sci Instrum 91(10):103502. ISSN: 0034-6748. https://doi.org/10.1063/5.0006672. eprint: https://pubs.aip.org/aip/rsi/article-pdf/doi/10.1063/5.0006672/14878456/103502/_1/_online.pdf
11. Granum P (2022) Measuring the properties of antihydrogen. Presented 19 Sep 2022. PhD thesis. Aarhus University. https://cds.cern.ch/record/2834233
12. Evetts N (2021) Solid-state nuclear magnetic resonance magnetometry at low temperature with application to antimatter gravity experiments by ALPHA. PhD thesis. University of British Columbia. https://doi.org/10.14288/1.0396355. https://open.library.ubc.ca/collections/ubctheses/24/items/1.0396355
13. Capra A et al (2017) Design of a radial TPC for antihydrogen gravity measurement with ALPHA-g. In: Proceedings of the 12th international conference on low energy antiproton physics (LEAP2016), p 011015
14. Smith G (2021) Characterization and analysis development for the ALPHA-g barrel scintillator. https://doi.org/10.14288/1.0401896. https://open.library.ubc.ca/collections/ubctheses/24/items/1.0401896
15. So C, Fajans J, Bertsche W (2020) The ALPHA-g antihydrogen gravity magnet system. IEEE Trans Appl Supercond 30(4):1–5. https://doi.org/10.1109/TASC.2020.2981272
16. Hodgkinson DL (2022) On the dynamics of adiabatically cooled antihydrogen in an octupole-based Ioffe-Pritchard magnetic trap. PhD thesis. The University of Manchester (United Kingdom)

Chapter 4
Hardware Developments

This chapter aims to provide a more detailed look at the internal workings of the ALPHA-g Penning trap electrodes and microwave systems. I will show development work undertaken for the hardware components in the ultra-high vacuum of ALPHA-g.

The development, preparation, and build of the Penning trap electrodes were a combined effort between myself and Dr. April Cridland. The design of the microwave system was led by Dr. Daniel Miranda Silvera with input from several ALPHA members including myself, while I also handled procurement, construction, assembly, and installation. Measurements of the microwave transmission and reflection were made by Professor Mike Hayden and Dr. Alberto Uribe Jimenez.

Installation of the UHV components was led by me and assisted by Dr. Chris Ørum Rasmussen, Dr. Alberto Uribe Jimenez, and Dr. April Cridland. Design of the installation procedures was handled by Dr. Timothy Friesen (Penning trap) and Dr. Daniel Miranda Silvera (microwave system).

4.1 ALPHA Penning Traps

As previously discussed, due to the magnetic field requirements for magnetic minimum trapping in Chap. 2, the inner radius of the electrodes in the antihydrogen confinement regions must be as close as possible to the windings of the octupole. To facilitate this the walls of the ALPHA thin electrodes are only 1.5 mm thick and electrodes are held together with polyamide tubes under tension with a thin aluminium wire through the tube. Each electrode is then electrically isolated from each neighbouring electrode by 8 ruby balls that maintain a 0.5 mm gap.

In 2018, the first iteration of the ALPHA-g trap was constructed and inserted. This followed construction methods previously employed in ALPHA-2 which uses an end electrode as in Fig. 4.1 to hold the polyamide tubes in a channel while the

A. M. W. Powell, *Magnetic Field Characterisation for Gravitational Free Fall Measurements of Antihydrogen in the ALPHA-g Experiment*, Springer Theses,
https://doi.org/10.1007/978-3-032-07190-3_4

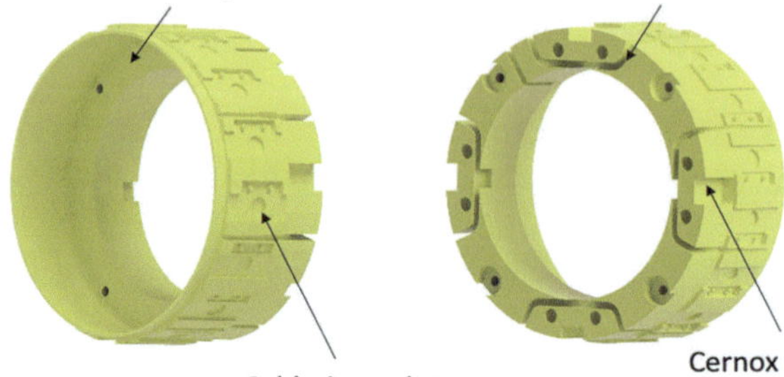

Fig. 4.1 End electrode designed to capture polyamide tubing and allow tensioning of ALPHA thin trap electrodes

tubes are individually pulled through to the other end of the electrode stack, held at tension, and crimped, then secured in divots in the opposite end electrode. This construction method leads to uneven tension applied to each polyamide tube and is a suspected cause of electrical shorts that have plagued ALPHA traps, making particle manipulations significantly more complicated and limiting the resolution of magnetometry field maps which will be discussed in Chap. 5. This construction method also requires the end electrode in Fig. 4.1 to have a radial step, this has implications for particle transfers between magnetic traps of ALPHA-g shown in Sect. 3.6. Given these issues it was felt an effort to improve the assembly of these thin electrodes could have a number of benefits.

In addition to the thin electrodes, which form the discussion for a majority of this chapter, a more typical Penning trap electrode assembly style having support rods screwed into a grounded ring at the end of the electrode stack is then used in the regions outside the antihydrogen trapping region; this allows for six way segmented electrodes that required for the particle manipulations described in Chap. 2. In ALPHA-g these more standard style electrodes, which we refer to as “Normal”, are at the bottom of the thin electrode assembly to allow for antihydrogen transfer into the upper magnetic traps.

4.1.1 Tension Method Development

After the removal of the first iteration of the trap in 2019, a development effort to make the thin trap assembly more robust was undertaken. In a series of iterations on the trap, we developed a new method of applying tension evenly to all the polyamide tubes.

The first design of a tensioning jig that would sit on the ends of the Penning trap during construction was 3D printed in polylactic acid (PLA). The concept was to

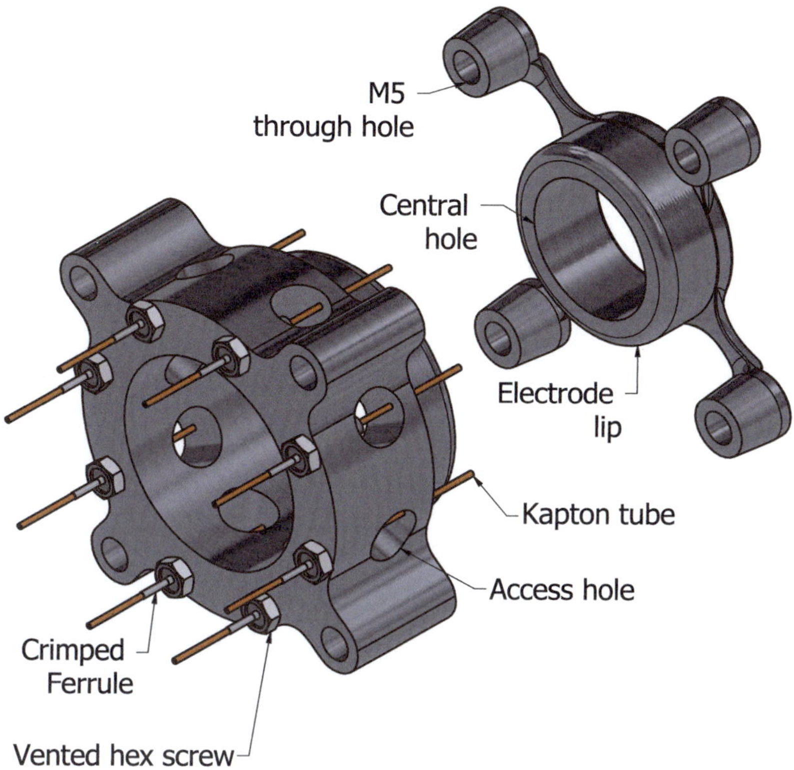

Fig. 4.2 Tension jig for bottom Penning trap final design. Figure credit A. Cridland

feed the polyamide tubes through vented screws in a jig, then crimp the ends of the tube to secure them in place. Once secure the vented screws could then be wound out, applying tension in a controlled fashion as can be seen in Fig. 4.2.

This proof of principle gave a rigid trap structure while the jigs were left attached but was not yet compatible with holding those tubes in place without the jigs and maintaining tension. A second iteration of the tensioning jig allowed the ends of the tube to be fixed to the electrodes with a thermally conducting epoxy, Stycast, and then the excess tubing to be removed. Initially we attempted to brush the Stycast onto the surface of the electrode and onto the polyamide tubes which proved somewhat effective but left the aluminium wire running through the tubing loose after the tension jigs were removed. As this technique improved and the confidence in the tension application was increasing we fixed this loose wire issue by cutting into the polyamide tubing with a scalpel and injecting the Stycast with a needle and syringe directly onto the wire in the tube. Following the gluing procedure and a 48 h curing time the tension jigs are removed from the electrodes by cutting the polyamide tubes at access holes on the side of the jig. A photograph of the ALPHA-g bottom Penning trap during the gluing procedure can be seen in Fig. 4.3.

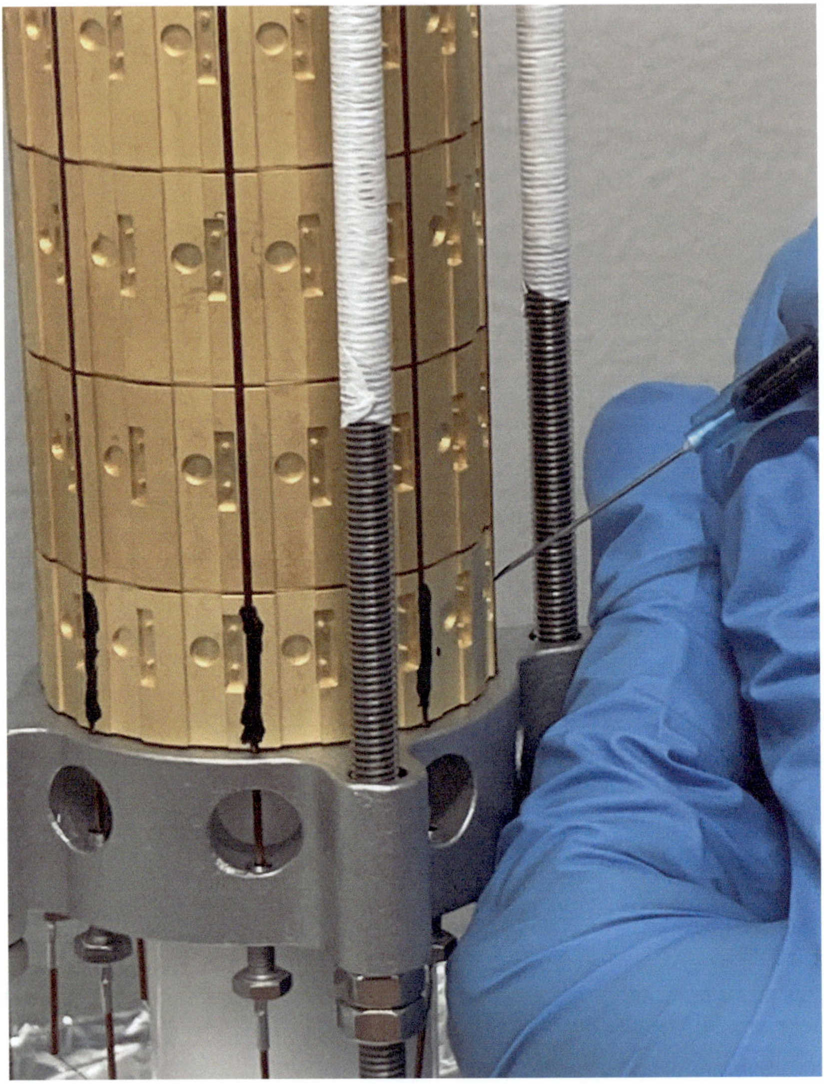

Fig. 4.3 Photograph of the ALPHA-g bottom Penning trap during the gluing process

4.1.2 ALPHA-g Bottom Penning Trap 2021

Figure 4.4 shows the ALPHA-g Penning trap assembly as installed in 2021. Details of the components below the catching gate (E1) will be discussed later in this chapter. Based on the developments in 2019 and 2020, it was decided that the trap for ALPHA-g would employ the new tension and glue method. However, to avoid a time consuming redesign of the electrodes connecting the thin and normal trap sections (E20),

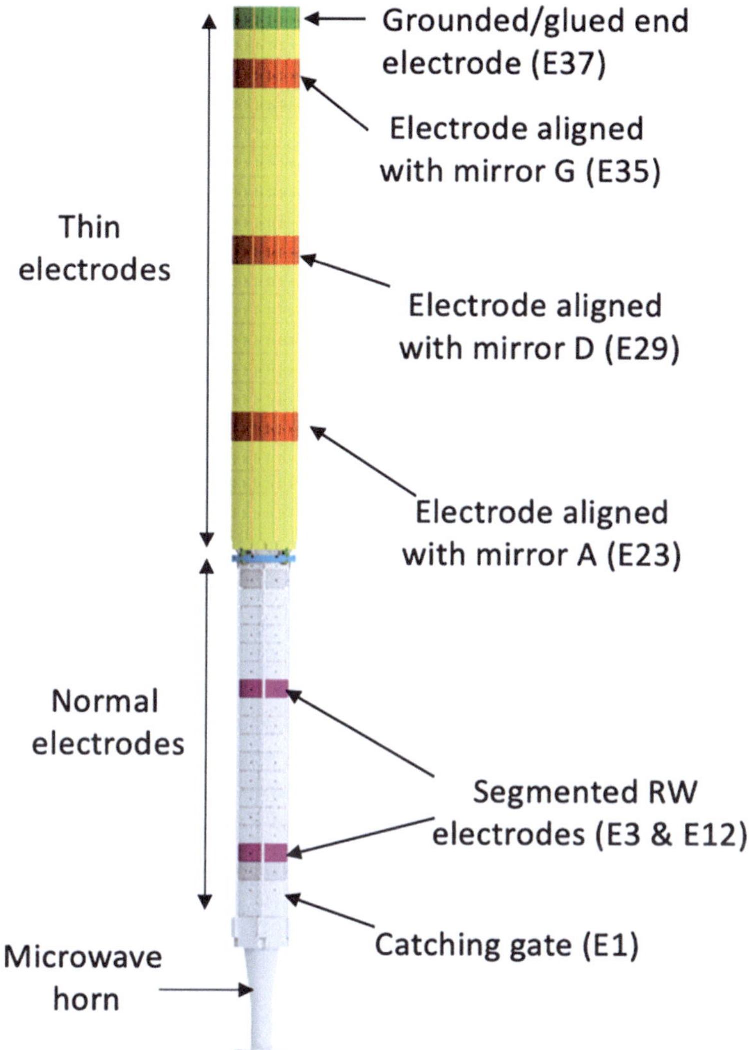

Fig. 4.4 Schematic of the ALPHA-g Penning trap as installed in 2021

only the top end of the trap (E37) would be glued. The final design for the jigs used in the bottom Penning trap assembly can be seen in Fig. 4.2. This version only applies tension from one side (E37) while maintaining the crimped hold of the tubes on the other end. This method only requires one electrode to be glued to hold the tension along the full length of the trap. This new approach did allow us to remove the radial step from E37 that featured in the 2018 trap as seen in Fig. 4.1. Simulations suggest the removal of this step could improve the transfer of antihydrogen from the bottom trap into the precision region by up to 30%. The transfer of antihydrogen to

Table 4.1 ALPHA-g bottom Penning trap electrode dimensions in 2021. All dimensions were manufactured with a general tolerance ±0.02 mm

	Electrode	Length (mm)	Inner radius (mm)	Wall thickness (mm)	Special features
Normal trap	1	28.44	14.80	3.00	Catching gate
	2	14.22	14.80	3.00	–
	3	14.22	14.80	3.00	Segmented (6 ways)
	4–11	14.22	14.80	3.00	–
	12	14.22	14.80	3.00	Segmented (6 ways)
	13–18	14.22	14.80	3.00	–
	19	23.00	14.80	3.00	Overlaps with E20
Thin trap	20	40.48	22.175	1.50	Radial step to connect traps
	21	16.00	22.175	1.50	–
	22 36	20.84	22.175	1.50	–
	37	16.00	22.175	1.50	Grounded, Glued

the precision region is expected to be inefficient due to the lower trap depth in this region, so gains in this transfer will benefit future precision measurements greatly. A summary of the electrodes in this trap, with dimensions and special features, can be seen in Table 4.1. The total length of the thin trap electrodes is 386 mm, while the normal electrodes sum to 274 mm.

4.1.3 *Soldering of ALPHA-g Thin Trap Electrodes*

Connecting the flexible circuit to the Penning trap while respecting the constraints of the inner bore of the cryostat is a challenge. The flexible circuit is well constrained by the electrode by channels cut into the electrode surfaces and laser cutting the flexible circuits to be held in these channels by a precision machined overhang. Due to the thin walls of the antihydrogen production region electrodes screws are not suitable and the circuits must be soldered in place. The most elegant solution to this issue has been to use a vapour reflow oven and solder all connections on the fully assembled trap simultaneously.

While the idea of soldering all points at the same time is appealing, it has practical challenges. First, aligning all solder points and maintaining some pressure during soldering to ensure a good electrical connection is essential. In previous iterations of the ALPHA traps, applying pressure during soldering has been achieved using a

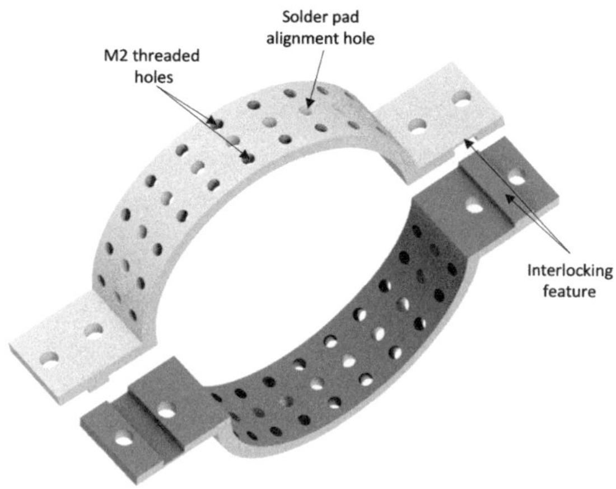

Fig. 4.5 Clamp used during soldering on the thin trap electrodes to apply even pressure around the joint and level the trap in the oven

simple aluminum ring around the electrodes with ad hoc threaded holes allowing a long screw to make contact with the flex. CERN have since replaced the reflow oven, it now has a smaller height and no longer fits the old configuration of aluminium rings. To get around this issue, each electrode now has a CNC machined soldering clamp as shown in Fig. 4.5. This clamp design allows four threaded M2 studs to act as feet and level the trap, allowing vapour flow to all surfaces and removing any uneven strains. There are then evenly spaced holes to allow for two threaded studs to apply pressure either side of the solder point, which is aligned over a through hole for visual confirmation the contact point is placed correctly.

While the soldering clamps appeared to have worked as designed several solder connections failed. It was later determined that as the flexible circuit had been produced in 2018, steps to remove oxide build up from the contacts should have been taken. All failed contacts were successfully soldered by hand using two soldering irons to locally heat the electrode and complete the solder connection.

4.1.4 ALPHA-g Top Penning Trap

The ability to measure magnetic fields in ALPHA-g is a must and the best methods for doing so rely on the Penning trap electrodes, as will be shown in detail in Chaps. 5 and 7. As ALPHA aims for a higher precision gravity measurement, the "precision region" of ALPHA-g shown in Sect. 3.6 will be utilized. The top Penning trap electrodes will be used for measurements of the magnetic field in this precision region through ECR magnetometry. Not only are these measurements necessary, they must be at a higher precision than magnetic field measurements that will be discussed in Chaps. 5 and 7 of this thesis. The previously discussed trap tension development have made it possible to now consider much longer Penning trap electrode stacks with new features that can allow access to the precision measurement region. Using

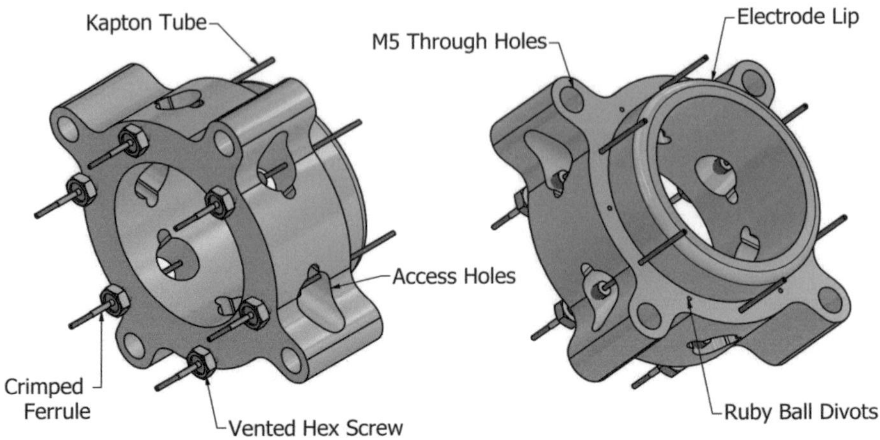

Fig. 4.6 Tension jig for top Penning trap final design. Figure credit A. Cridland

an adapted set of jigs in Fig. 4.6 both ends of the trap will have the polyamide tubes glued in place. This improved tension allows a reduction in the number of polyamide tubes required, potentially making space available on the electrode surface for more electrical signals or other features. As such, this top trap design has reduced the number of polyamide tubes from 8 to 6 to test the long term stability of an electrode stack with this reduced tension wire configuration.

While the limitation of tension has been removed, a second limit to the length possible was set by the number of electrical signals that could be cabled out from the top of the cryostat. At the time of this trap design, only three 51 pin Micro-D Glenair feedthroughs were available giving a 18 electrical signals per connector, allowing for each to have shielding with grounded pins, leading to a total of 54 independent signals without a major redesign of the vacuum system. Including two rotating wall electrodes with six sections, therefore requiring 12 signal lines, leaves 42 as a maximum number of electrodes. To maintain the resolution of the magnetic field maps electrodes cannot be too long and 42 was not enough to cover the entire region from the top capture solenoid to the bottom of the precision region.

Given the necessity of ECR measurements for any future gravity measurements in the precision region of ALPHA-g, the decision was made to prioritise the placement of electrodes in the precision region. The design for this top Penning trap renders the top magnetic trapping region unusable for antihydrogen confinement. The trap increases the length of the copper rods heat sinking the trap to the cyrostat compared to the bottom Penning trap, has a largely unchanged normal electrode assembly compared to the bottom trap, and a 641 mm long thin trap region extending from mirror G top to below the bottom precision coil, this can be seen in Fig. 4.7. Careful attention was paid to align electrodes under the two precision coils for the best magnetometry resolution possible.

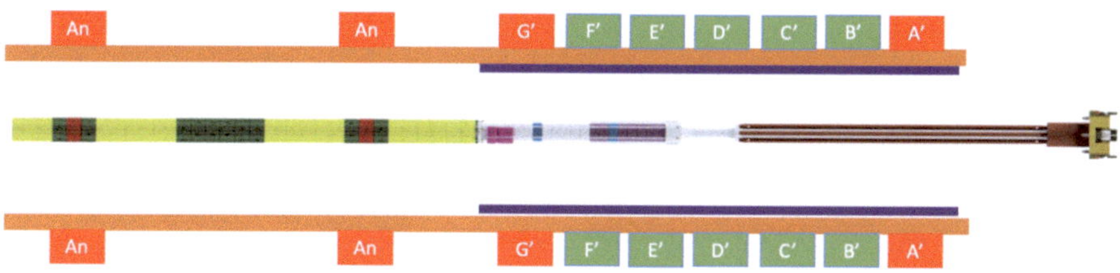

Fig. 4.7 Cartoon of top penning trap alignment with internal magnets. Not to scale

Some details about the planned trap are shown in Fig. 4.8, there will be a region used to test a colloidal graphite coating (Aquadag E Colloidal Graphite 12661 with Colloidal graphite alcohol base 12660 from Electron Microscopy Sciences) and how this impacts plasma preparations, this region is shown in purple. To make the machining of the thin electrodes simpler and the assembly process for the full set of thin electrodes easier, the feature that captured the flexible circuits in channels on the electrodes were removed. The channels remain and the flexible circuit is designed to sit in the channels and be constrained by a clamp system machined from polyether ether ketone (PEEK). The PEEK is compatible with cryogenic temperatures and ultrahigh vacuum while electrically insulating. In Fig. 4.8 these clamps are shown in pink and are screwed in to the normal trap electrodes. The electrodes highlighted in red are positioned to align with the design center of the precision region coils and will be key in mapping the magnetic field in this region. The electrodes in green either side of this aligned electrode are short to allow for precise manipulations of the electrostatic well minimum used to position electrons during mapping, see Chap. 5 for details. There is another region of short electrodes in the middle of the thin trap, this is to allow for plasma manipulations and extraction of target plasmas as in Sect. 5.1.1.

The design for the top trap was concluded while technical issues around the soldering and insertion of the trap were still unresolved. The soldering issue is related to the length of the thin trap in this configuration, it is too long for the reflow oven to solder flexible circuits to previous thin trap electrodes and discussed in Sect. 4.1.3. The solution to this issue was for each electrode connection to be soldered by hand with the electrode locally heated with a soldering iron.

The insertion of the trap into the cryostat is an issue because it must be installed with the bottom Penning trap electrodes in place. This reduces access from the opposite side, which for previous trap insertions has allowed the trap to be safely constrained to avoid collisions with the narrowest point of the cryostat. Given this additional complication an insertion and alignment frame, including a winch for controlled lowering of the trap, was designed by Reece Stefanyshyn. The top trap was successfully installed in early summer 2024.

A summary of key differences between the traps is shown in Table 4.2.

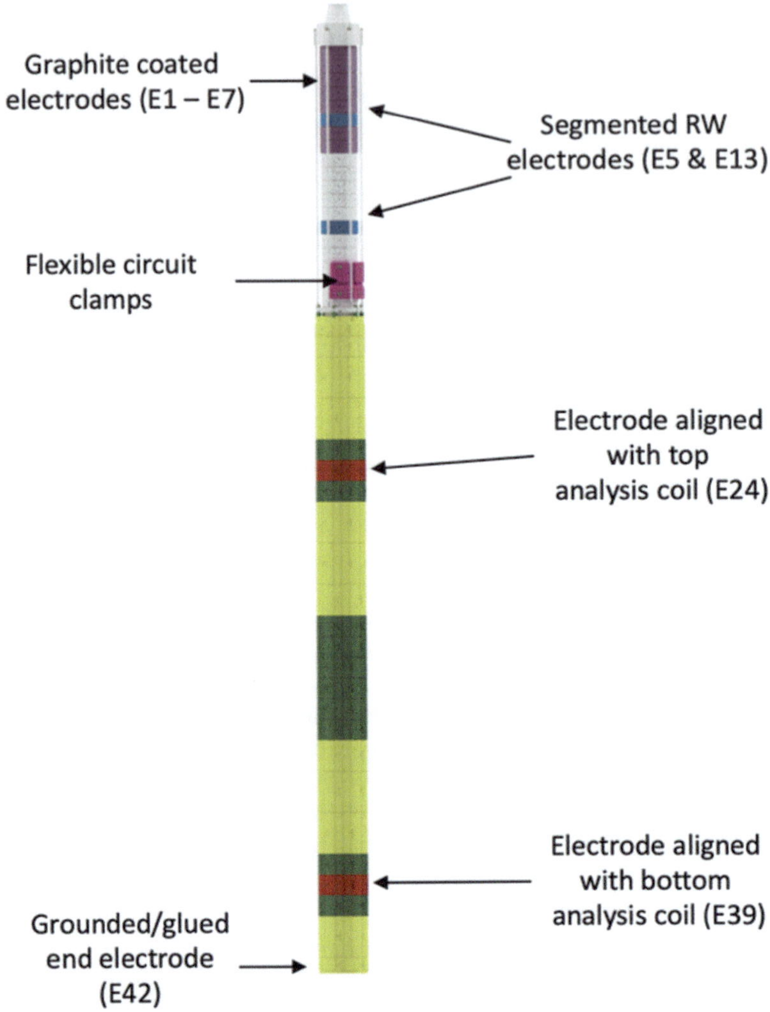

Fig. 4.8 Detailed view of top penning trap design features

Table 4.2 A comparison of the ALPHA-g Penning traps

Trap	Electrodes	Length total 4 K to end electrode (mm)	Length thin trap (mm)	Length normal trap (mm)	Surface coating
Bottom	37	852	386	274	Gold plating
Top	42	1506	641	274	Gold plating + graphite

4.2 ALPHA-g Microwave Injection

In the ALPHA-2 apparatus, a waveguide transmits microwaves to the trap through the vacuum system, increasing power reaching the trap. For ALPHA-g the space constraints and length of the machine made an equivalent design challenging. The subsections that follow describe the design choices that were made to allow the efficient injection of the broadest frequency range possible to enter the trap. This broad range is desirable as one considers experiments at different magnetic field strengths, this could be for decreased sensitivity to field errors or to allow for extrapolation to zero field. In ALPHA-g, high frequency propagation is also vital to allow for higher magnetic field measurements with the ECR technique, this will be discussed extensively in Chaps. 5 and 7.

4.2.1 Insertion Path Design

The microwave pathway for ALPHA-g begins with a Keysight E8257D synthesizer. This has a frequency range of 100 kHz to 50 GHz and a frequency resolution of 0.01 Hz. The output of this synthesizer is a 2.4 mm coaxial connection to a 4 m long low-loss DC to 40 GHz coaxial cable (Conduct RF: PFA89-C1C1-4M0). While this cable is technically rated to 40 GHz only, the losses were sufficiently low up to 49 GHz for use in our apparatus.

The pathway continues with an adapter from 2.4 mm coaxial cable to WR42 waveguide. This adapter is connected to a quartz vacuum window, sealing the ultra-high vacuum of the apparatus from atmosphere while allowing the transmission of microwaves. The window is mounted in a flange connected to rigid WR42 waveguide that has a bend with a CF16 vacuum flange brazed on the other end as seen in Fig. 4.9. Internally, a full set of custom waveguide sections were designed to minimise heat leaks from the trap through the waveguide while transmitting the correct range of microwaves with acceptable losses. The first section in Fig. 4.10, has a connection to the CF16 flange and a short section of flexible beryllium copper WR42 waveg-

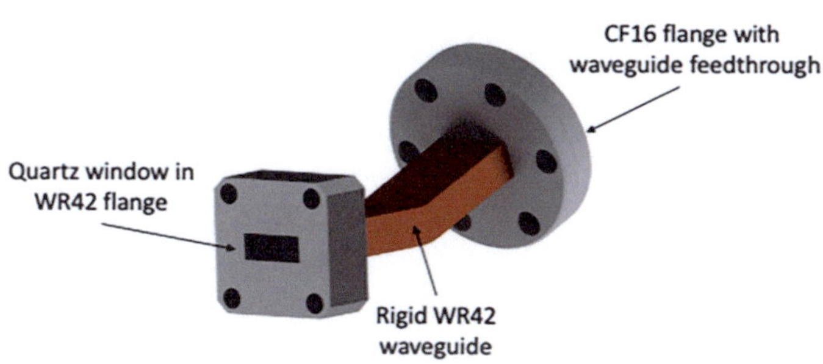

Fig. 4.9 Quartz microwave window assembly to seal ultra high vacuum from air assembly

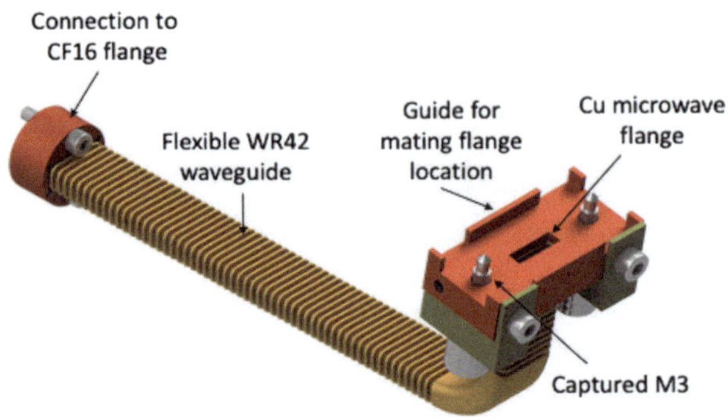

Fig. 4.10 Section 1 of the waveguide path. This flexible guide runs from the microwave window assembly to the titanium heat break

uide. This small circular connection allows installation of this section through tight constrictions and connection must be made before the CF16 vacuum flange for the window section (Fig. 4.9) is sealed. The final, and most critical component in this section is the custom microwave flange, it has edges and captured screws to allow connection in the vacuum chamber with one hand. This feature was necessary due to the tight space constraints below ALPHA-g inside the bellows connecting the interconnect to the ALPHA-g UHV.

Section 2 of the assembly begins with a titanium flange to act as a heat break and decrease the thermal load on the cryostat. It is again connected to flexible waveguide and the other end is clampled at the 50 K stage in Fig. 4.11. The final section,

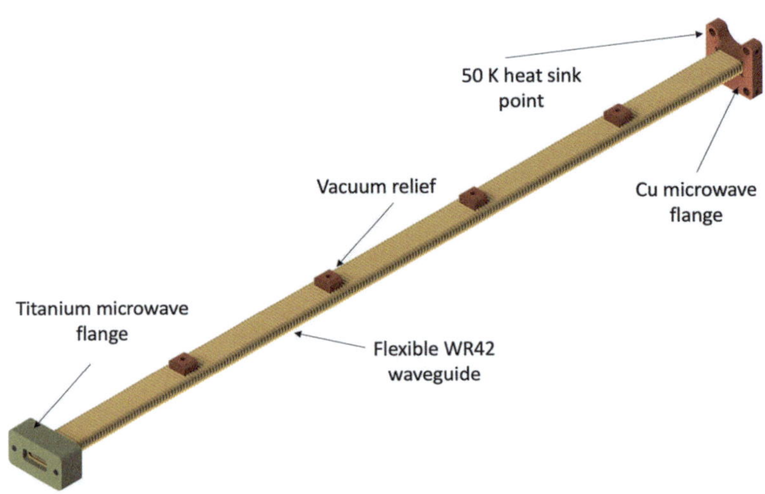

Fig. 4.11 Section 2 of the waveguide path. This section between room temperature and 50 K stage with a Titanium heat break

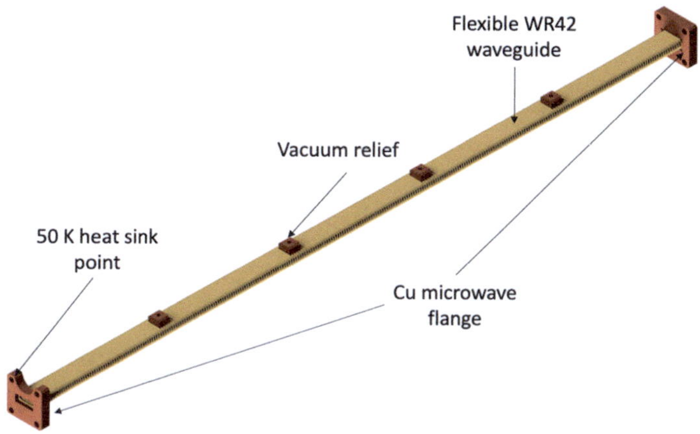

Fig. 4.12 Section 3 of the waveguide path. This section is between 50 K and 4 K stages

Fig. 4.12 is connected to the 50 K stage and the trap cup assembly at 4 K which couples the microwaves into the trap by a waveguide braised into a copper beampipe. The full assembly with its connection to the trap can be seen in Fig. 4.13.

To overcome several technical issues the waveguides were soldered to the microwave flanges by hand with silver solder. The face of the flange sat on a hot plate and solder was applied from above once the waveguide was positioned in the flange cut-out as shown in Fig. 4.14. This proved very successful.

The next stage was to provide vacuum relief to the waveguide. This had to be done while minimising impact on the microwave properties. Copper tabs with a guide hole were soldered to the outside of the waveguide, again using a hot plate and silver solder, in preparation for drilling relief holes. To avoid burrs on the inner surface an electrical discharge machining (EDM) drill with a water dielectric was used, this ensured the waveguide could be cleaned and made vacuum compatible compared with an oil dielectric. The key parameter in this process is the aspect ratio of the hole, the hole diameter must be smaller than the microwave wavelength and the depth of the hole produced by the vacuum relief guide ensures the microwave field has decayed and there is no leakage.

To improve the transmission of microwaves into the trap there is a smooth transition from the 16 mm diameter beam pipe required to allow the copper connecting rods to hold the trap to the inner diameter of the electrodes at 27 mm. This section, shown in Fig. 4.15, was made wider than previous designs. To increase the range of transmittable frequencies, with the low frequency cut-off given by,

$$f_{\text{cutoff}} = \frac{1.8412c}{2\pi r} \tag{4.1}$$

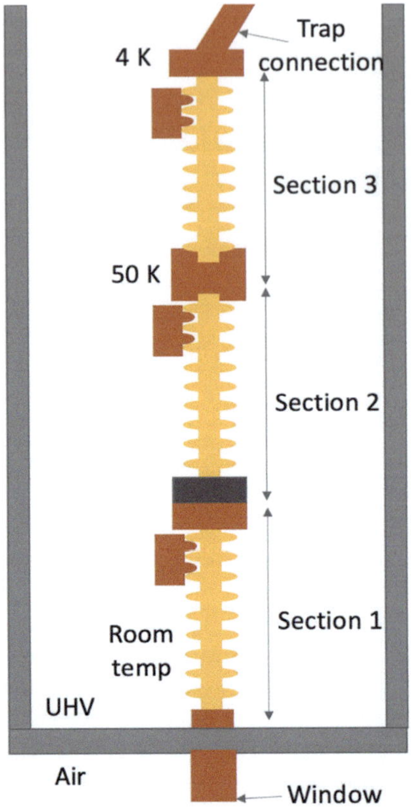

Fig. 4.13 A cartoon to show how the various sections of waveguide form the full path from outside the UHV into the trap

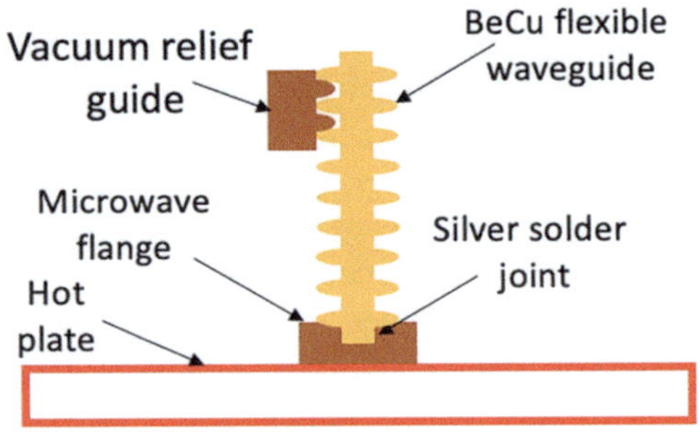

Fig. 4.14 A cartoon showing the method used to solder microwave flanges to flexible waveguides

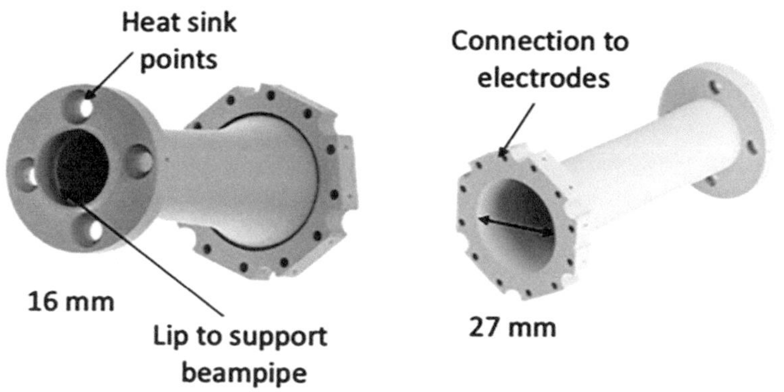

Fig. 4.15 Smooth section leading into the Penning trap transitioning from a 16 mm diameter to 27 mm

with c as the speed of light in the medium and r as the radius of the waveguide. The smallest inner radius was increased from 6 mm ($f_{\text{cutoff}} = 14.6\,\text{GHz}$) to 8 mm ($f_{\text{cutoff}} = 10.9\,\text{GHz}$). While this is not necessary for the experiments discussed in Chap. 6, it is interesting when considering future free fall or spectroscopy experiments that can benefit from operation at a lower magnetic field.

To confirm transmission through the microwave window and into the trap, measurements of the transmission and reflection were undertaken. Using a Keysight N5224B 10 MHz to 43.5 GHz vector network analyser (VNA), which was calibrated with calibration kit Keysight Ecal N4693D, successful transmission of microwaves was observed.

Following these changes to the internal UHV components of ALPHA-g we had a system ready for experiments with antihydrogen and charged particles. The work in this chapter was largely motivated by the need to perform ECR magnetometry at the maxima of mirrors A and G. The next chapter will introduce the ECR technique and show the first ECR measurements in ALPHA-g.

Chapter 5
Electron Cyclotron Resonance (ECR) Magnetometry

The previously described neutral and Penning trapping systems in ALPHA are not conducive to traditional magnetic field measurement methods, such as nuclear magnetic resonance (NMR) or Hall probes, due to the combination of an ultrahigh vacuum and operation at cryogenic temperatures. However, given the area of interest is within a Penning trap, charged particles offer an alternative method of field measurements.

The basis of the ALPHA ECR technique is the cyclotron motion of charged particles, in this case electrons, in a magnetic field as described in Sect. 2.9. If the cyclotron frequency of an electron at a point in a magnetic field can be found, the magnetic field in that location can be calculated as,

$$B = \frac{2\pi m_e f_c}{q}. \tag{5.1}$$

Where B is the magnitude of the magnetic field, f_c is the cyclotron frequency, m_e is the mass of the electron, q is the elementary charge. The cyclotron frequency for a cloud of electrons can be determined by irradiating the cloud with microwaves, if the microwave frequency is resonant with the cyclotron frequency, heating of the cloud will be observed. This chapter will give background on the reservoir electron cyclotron resonance (ECR) magnetometry technique, including the preparation of an electron reservoir, target plasmas, parameters of the target plasmas and diagnostics used to determine these characteristics. I will then go on to show the methods for identifying the ECR frequency, mapping magnetic fields and the developments that have allowed for the first magnetic field mapping in a highly in-homogeneous field using this technique.

A. M. W. Powell, *Magnetic Field Characterisation for Gravitational Free Fall Measurements of Antihydrogen in the ALPHA-g Experiment*, Springer Theses,
https://doi.org/10.1007/978-3-032-07190-3_5

5.1 Reservoir ECR Method

Magnetic and electric field diagnostics using the non-destructive measurement of the quadrupole normal mode (an oscillation in the aspect ratio of the plasma, also known as a breathing mode), as a proxy for cyclotron heating, of a large electron plasma driven by resonant microwaves were used in ALPHA for several years [1]. While this tool provided in situ diagnostics in ALPHA it was precision limited by the size of the plasma (due to large magnetic field gradients) and repetition rate of measurements when trying to follow drifting magnetic fields. To counter these issues, ALPHA collaborators at University of California Berkeley developed a method for extracting target electron plasmas from a larger electron reservoir [2]. This technique has been adapted for use in the ALPHA Penning traps and is implemented as follows.

5.1.1 Target Plasma Preparation

A plasma with typically 10–40 million electrons and a radius of approximately 1 mm, called a reservoir, is prepared following the electron loading and SDREVC process described in Sect. 2.3.3. An example of SDREVC potentials for reservoir preparation in ALPHA-g is shown in Fig. 2.2. The reservoir is prepared in an approximately 0.9 T magnetic field compared to the standard 2–3 Tesla used to develop SDREVC which increase cyclotron cooling rates. This is due to the antihydrogen release protocol requiring the capture solenoid magnetic field to be removed to not bias the annihilation position, this will be discussed in detail in the next chapter. The ability to load electrons and perform SDREVC in this field configuration was vital to successful measurements in ALPHA-g and required extensive tuning of the reservoir plasma manipulation.

Following SDREVC, the reservoir is stored in a deep electrostatic well at the bottom of the trap (electrodes 2–4 of the normal trap described in Chap. 4) and the process of target extraction can begin, this can be seen in Fig. 5.1. To summarise, two electrodes above the reservoir holding well are set to a positive potential to elongate the reservoir. Once stretched out, the nearest electrode to the original reservoir well is set to zero potential, pinching off a small number of electrons in the second electrode, forming what I will henceforth call the target plasma. The target is manipulated into the region of the trap to be measured and the process is then repeated many times to remove many target plasmas, typically 100–200. The key parameter to tune in this process is the potential of the electrodes involved in the "pinch" (electrodes 5 and 6) relative to the reservoir. It should be set such that the difference in potential from the maxima of the reservoir well and the new electrode potentials match the amount of space charge in the reservoir. It then follows that this potential must be adjusted as charge is removed from the reservoir and it expands radially, as such the potentials get progressively more positive.

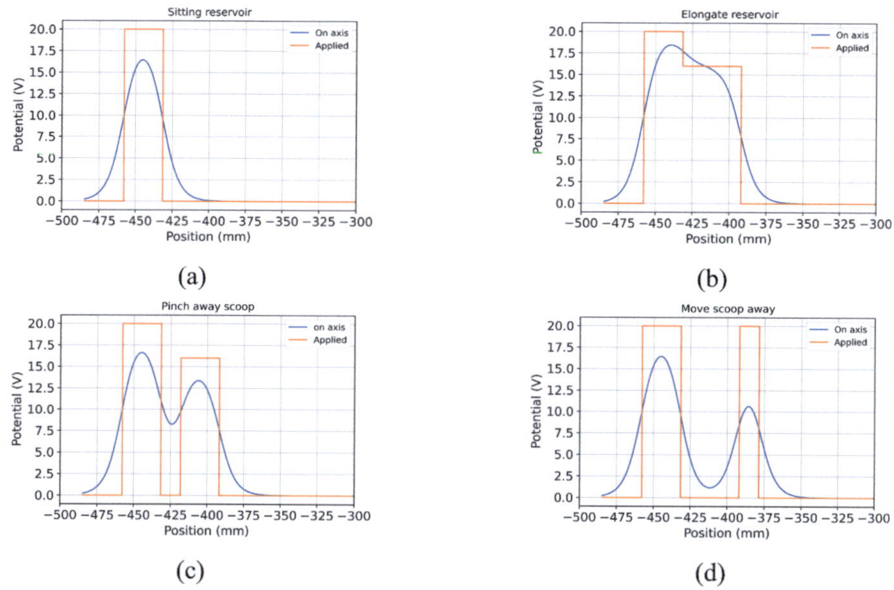

Fig. 5.1 Example sequence of potentials used to extract a target plasma from an electron reservoir. **a** the reservoir is ready for target extraction, **b** the reservoir is elongated, **c** the target plasma is pinched from the reservoir and **d** the target is moved to the measurement location

Once the target electrons are removed from the reservoir they can be moved to the location of interest by electric potential manipulations. The number of target plasmas that can be removed from a reservoir is variable and often depends on the stability of the reservoir. As the targets are removed the reservoir becomes hollow and develops a diocotron motion (a plasma instability). This in turn causes an increasing diocotron motion in the extracted targets, making them less usable as target number increases, particularly if this instability is great enough that target plasmas miss the MCP when released for imaging or temperature measurements.

5.1.2 Evaporative Cooling of Target Plasmas

As a brief aside, after the extraction of the target it is possible to evaporatively cool following the same process as has been previously described with larger plasmas. While this is an effective method of cooling the target sample and reducing the overall electron number without exhaustive tuning of the extraction potentials, but there are compromises. Firstly, the evaporative cooling (EVC) radially expands the plasmas as previously discussed in Chap. 2. This radial expansion was not a considerable issue in the Berkeley experiment as they worked in a single solenoid field, ALPHA however has a considerable radial field gradient due to use of octupole magnets. The second, and more considerable issue, was the time taken for the EVC process. Due to the slow manipulations required and the hundreds of times these manipulations occur this increases the overall time for an ECR measurement considerably. When one considers machine time is a limited resource and many ECR measurements are required to fully understand the magnetic environment, added time to each measurement is highly detrimental to the overall success of the campaign. The increased time also makes the characterisation of fast changing magnetic fields even more challenging. As such, tuning of electron numbers was done solely by tuning the extraction potentials shown in Fig. 5.1 until small and consistent target plasmas are produced and the required precision of each measurement can be achieved.

5.1.3 *Target Plasma Parameters*

MCP images of the targets are taken to estimate the radial size and number of particles, as well as allowing observation of any diocotron instabilities. An example of a basic MCP analysis used when tuning the extraction potential level is seen in Fig. 5.2. In this analysis, pixels are summed in either the x or y axis and a Gaussian fit to the peak in that dimension. From this an estimate of relative target plasma size and intensity is possible and outputs of this analysis type are shown in Fig. 5.3. The standard analysis used by ALPHA [3] is slow and the fitting procedure often misses the small

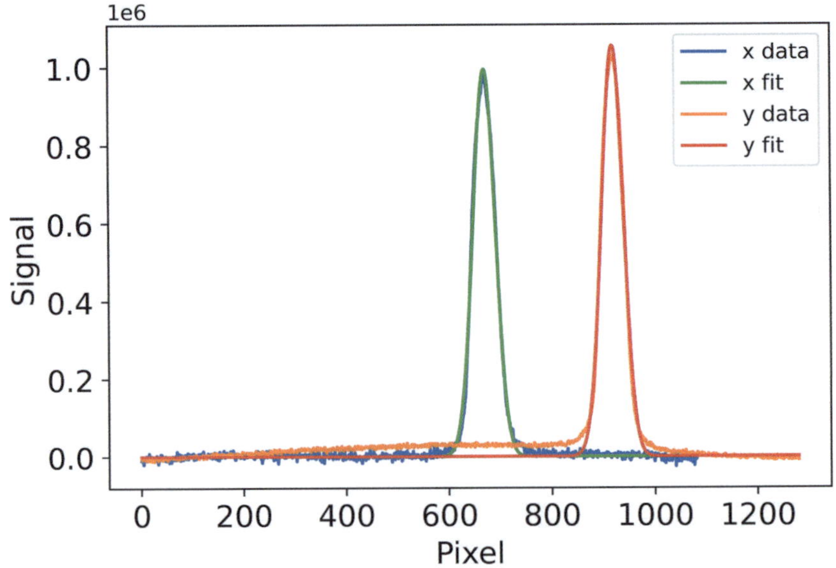

Fig. 5.2 An example of a rapid analysis method for relative size, charge amount, and extraction position of target plasmas

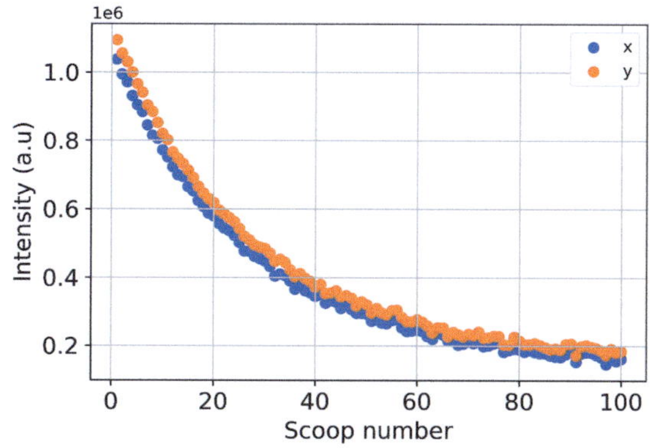

(a)

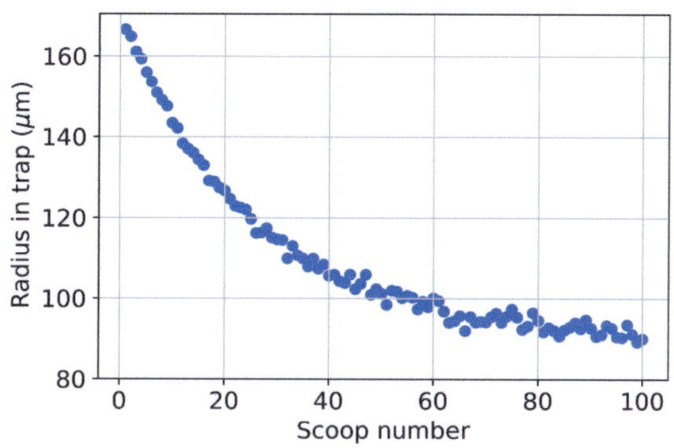

(b)

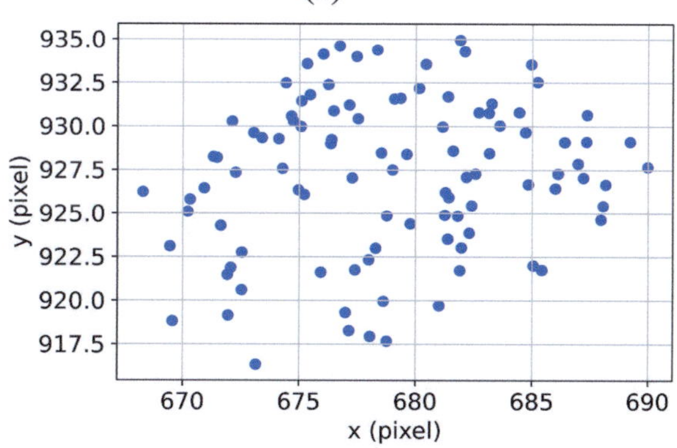

(c)

Fig. 5.3 Plasma parameters from MCP images of target plasmas used in ALPHA-g. **a** the changing number of electrons per target plasma **b** shows the changing radius per target plasma and **c** extracts a position to show any diocotron instabilities amounts of charge in the targets. This makes it inefficient for target plasma tuning and therefore it is preferable to use the less sophisticated but faster method to find relative parameters for each target plasma described above. The targets used in the collection of data in this thesis have a radial extent of 0.1 mm inside the trap and have 10^4 electrons, though these values do vary.

As discussed in Chap. 2, a collection of charged particles must satisfy certain criteria to be described as non-neutral plasmas. The length and density of these target plasmas is on the boundary of the plasma regime and as such there are collective behaviours but the target plasma does not completely flatten the electric potential of the confining well.

5.1.4 Microwave Irradiation and Heating

Now that a target plasma has been prepared and moved into the target position, microwaves are injected into the Penning trap as laid out in Sect. 4.2. Electrons in the Penning trap perform simple harmonic motion and addition of microwaves can be therefore allow the electrons to be treated as a driven Lorentz harmonic oscillator [2, 4]. Microwaves on resonance with the cyclotron frequency will add energy to the system, heating the target plasma. Microwaves propagate through the trap as though in a cylindrical waveguide. As the frequencies of interest have a smaller wavelength than the diameter of the trap, the system is over-moded and a complicated standing wave pattern forms in the trap. This leads to variable microwave power as a function of both position and frequency and as such discussion in this section will be qualitative and trying to motivate the heating rather than as a basis for modelling the electron response in the ALPHA traps during irradiation.

For now ignoring the caveats above and assuming the microwaves propagate as a circularly polarised plane wave wave along the trap axis with an electric field given by

$$\vec{E} = [E_-(\hat{x} - i\hat{y}) + E_+(\hat{x} + i\hat{y})]e^{i\omega t}. \tag{5.2}$$

The E_+ term is non-resonant so can be ignored and power absorption can be found to be [2],

$$P(\omega) = \frac{e^2\tau}{m_e}\frac{|E_-|^2}{1 + \tau(\omega_c - \omega)^2}, \tag{5.3}$$

where e is the elementary charge, m_e is the electron mass, ω_c is the cyclotron frequency, ω is the drive microwave frequency. The final term, τ, is a decoherence term used to introduce damping to the system to account for collisions of electrons within the target and with background gases. The shorter this time the broader the resonant structure, therefore offering lower precision. The pulse length of the microwaves used in the experiment are longer than the expected decoherence time scale, such that the decoherence no longer dominates the line shape.

As the electrons are confined in an electrostatic well and have a bounce frequency as they traverse the well, the microwave electric field seen by the electrons has a spatial dependence,

$$E_{-}(z,t) = E_{-}[e^{i\omega t}e^{-ikz_0 sin(\omega_z t)}]. \tag{5.4}$$

By examining Eqs. 5.3 and 5.4, the condition for heating is satisfied when $\omega = \omega_{c'} + m\omega_z$, where m is an integer, indicating heating will occur at not only the ECR frequency but with side peaks at multiples of the bounce frequency.

5.1.5 ECR Peak Identification

Given that heating of the electron target plasmas can occur at $\omega_{c'}$ and $\omega_{c'} + m\omega_z$, definitive statements about the field can only be made once it is verified that the heating corresponds to a peak from the true cyclotron frequency. The peaks related to the bounce frequency, at $m\omega_z$, can be most easily identified by their shape. It is usually seen as a broad (many MHz wide) peak that has a very sharp asymmetric side, with the sharp edge indicating the true bounce frequency of the well and the broad side showing that due to the plasma space charge flattening the potential and producing a distribution of bounce frequencies.

In some cases the electrostatic well used to confine the electrons in specific locations in the Penning trap will obscure this shape significantly, long asymmetric wells are sometimes required due to the trap electrode lengths. In the instance where the peaks cannot be easily identified by shape and the solution is to vary the depth of the electrostatic well while maintaining the axial position of the minimum. As the depth of the well changes, the bounce frequency changes but the position of the minimum is maintained, therefore the target plasmas position in the magnetic field is unchanged and thus the cyclotron peak will not move but the side peaks will shift. An example of the peak identification procedure can be seen in Fig. 5.4. In this example the confining potential for the blue spectrum is deeper, and therefore the axial bounce frequency is higher.

There is a new question of peak identification when one explores higher resolution scans related to peaks corresponding to rotation at $\pm n\omega_R$. For the studies in this thesis these are not relevant, as the precision requirement is not high enough to require the resolution of the rotational sidebands, but a guide for solving this issue has been provided by Hunter et al. [2].

5.1.6 Modified Electron Heating Estimation

Due to a repeated issue with the SIPM signal from the ALPHA-g apparatus, the temperature calculation routine discussed in Sect. 2.3.1 would fail for the target plasmas. As a result, the temperature of the hottest plasmas would often be underestimated,

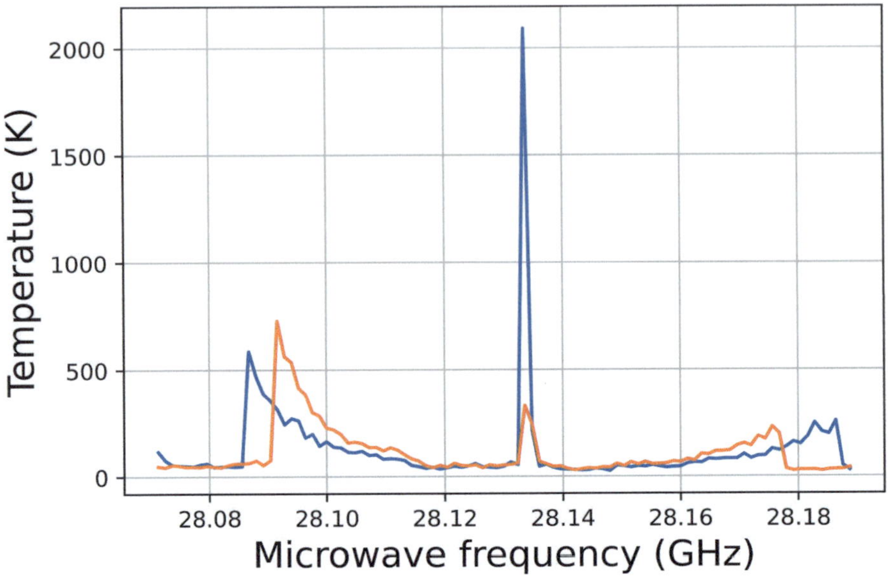

Fig. 5.4 Example ECR spectra with 100 target plasmas each at the same axial location of the ALPHA-g Penning trap with different well depths

leading to a dip in the peak structure near the maxima and broadening any attempts to fit the peak as can be seen in Fig. 5.5. To work around this issue I implemented a basic estimation of relative temperatures based on counting the number of time bins the SIPM signal is above a background value. The hotter the electron plasma the wider the time spread of the signal during the release. Figure 5.6 shows three target plasma extractions to the SiPM at different temperatures after microwave irradiation. In the Fig. 5.6a, the plasma is cold and far from the ECR resonance, the Fig. 5.6b example is on a rising side of the resonance peak, and Fig. 5.6c is near the maxima. It is clear from Fig. 5.6c signal that fitting an exponential is impossible or at the very least, meaningless. The relative approximation then is to use the end of the trace, coloured in green, and average to find a background level. I have then set the above background threshold as 3.5 times above the background to remove noise, the counts that do not meet this criteria are then in blue with the signal in orange. The number of time bins in the orange region is then related to the temperature as higher energy particles can escape the blocking potential well earlier.

It should be noted that this method is not ideal and is particularly sensitive to changing number of electrons in the target plasma. This new heating estimation was only intended to improve the quality of the current data set and not as a permanent change for the ALPHA ECR technique. All remaining results in this thesis will use this heating estimation unless explicitly stated otherwise to ensure the same analysis was performed on all data.

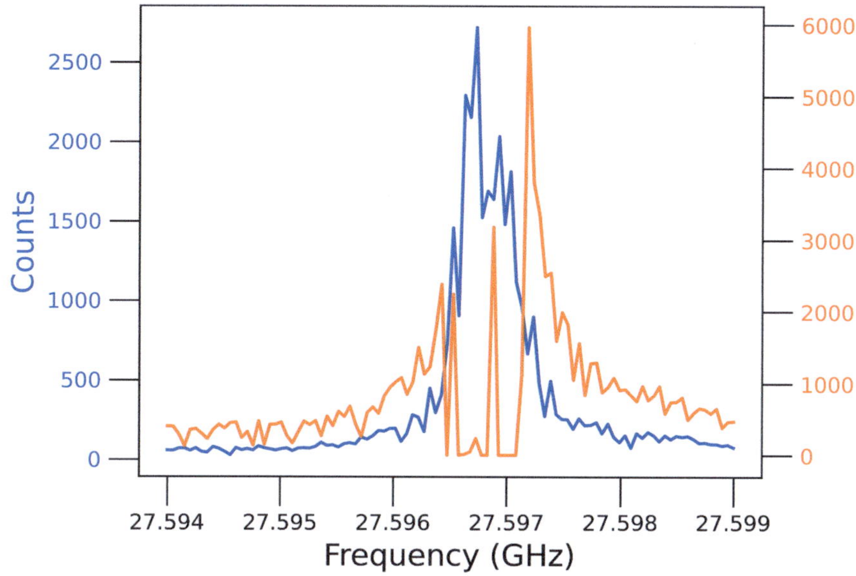

Fig. 5.5 Comparison of two different heating estimation techniques showing the issue arising from poor fitting

5.2 ECR Technique Development in ALPHA-g

After the initial commissioning of ALPHA-g the following data were taken as part of a development program in preparation of a gravity measurement. There were still several hardware and software changes being made to the overall system while this work was being carried out and I will outline any caveats to measurements as appropriate.

This section begins with a number of magnetic field maps along the axis of the Penning trap with different magnetic fields applied. It then moves on to discuss the key development of measurements in a high magnetic field. Finally it will show how measurements are made in multiple locations simultaneously to reduce sensitivity to decaying fields.

The process of mapping the magnetic field requires the target plasmas to be confined in different positions along the trap axis. This is done through calculation of the electrostatic potential with different applied voltages to each electrode. The minima of the well is taken to be the confinement position of the electrons and given the small particle number this is a reasonable assumption as the potential will not be flattened by the space charge of the target plasma.

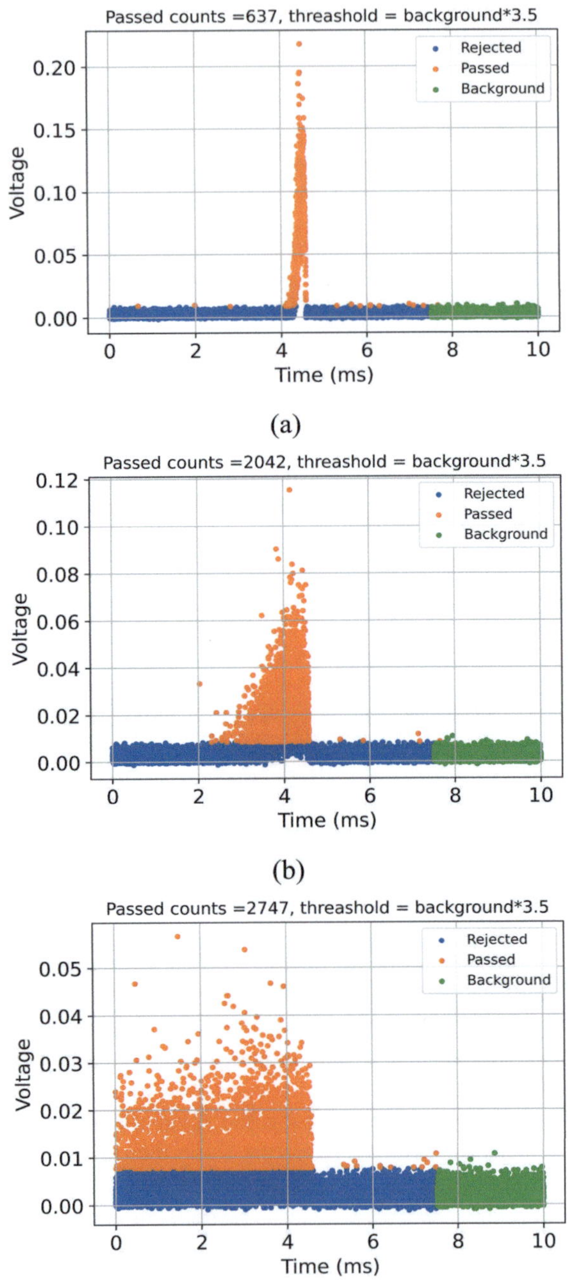

(c)

Fig. 5.6 Example SIPM signals for three target plasmas with different temperatures after microwave irradiation and the implementation of a crude relative temperature estimation for these data

5.2.1 *Persistent Fields*

As it is an important motivation for many of the measurements and techniques about to be discussed I will take a brief aside to discuss persistent fields. This is a phenomenon by which superconducting material in a magnetic field can have induced currents form in the presence of a change in magnetic field. These currents, which decay over time, produce magnetic fields. The induced currents (and therefore the induced magnetic fields) can decay with varying decay time constants that must be experimentally determined. These persistent fields and their decays can be large in the ALPHA apparatus depending on the magnetic field change that induced them and so must be carefully characterised for experiments where a major systematic component is the magnetic field such as will be shown in Chap. 6. The designs of the ALPHA-g magnet system as shown in Sect. 3.6 is intended to minimise the magnitude of these persistent fields by having minimal superconducting material but most critically aims to minimise asymmetries of the persistent fields by having material placed symmetrically.

5.2.2 *Axial Mapping of the Babcock Magnet*

In preparation for a gravity measurement the magnetic field without the mirror coils A and G was studied extensively to ensure there are no stray fields on the order of the gravitational potential difference across the magnetic trap that could bias annihilation distributions.

The first point of study was the 1 T field from the Babcock magnet, required for charged particle confinement, this was done for several important reasons. First, mapping the magnetic field produced by the Babcock in the region of the top and bottom mirror coils to look for asymmetries is vital. Any asymmetries must be compensated for experimentally and in simulations using current in the mirror coils later. The first attempts at this used the field from the capture solenoid to prepare the reservoir and beamline magnets energised when extracting plasmas to the MCP. This gave an indication of the fields but has gradients of several Gauss between the top and bottom mirror coils. This was estimated to be on the order of 4 Gauss and was later confirmed by measurements.

As the reservoir became more developed, loading and plasma preparation without the high magnetic field produced by the capture solenoid was possible. After the coaxial alignment of the Babcock magnet relative to the Penning trap electrodes was adjusted it was possible to extract plasmas without beamline magnets. Details of the alignment campaign using electron plasmas are outside the scope of this thesis and will be covered by Andrew Christensen of University of California Berkeley using techniques described elsewhere [5, 6]. The major outcome of the alignment campaign was the first magnetic field map with no stray fields from the beamline

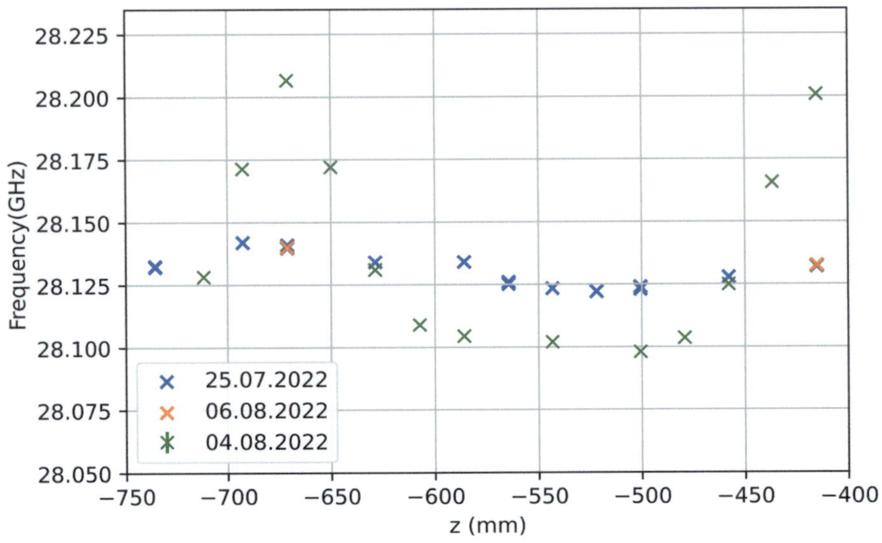

Fig. 5.7 An on axis field map of the Babcock magnetic field showing variation in shape due to induced fields from testing of mirror coils A and G

magnets or the capture solenoid and can be seen in Fig. 5.7. Unfortunately, during this period the remaining internal superconducting magnet circuits were undergoing commissioning and testing of quench protection circuits. This led to variation between measurements depending on the magnet ramp sequence in the hours and days before a measurements, making controlled experiments difficult. After the completion of magnet testing, the first "clean" maps of the background field were made and showed an offset in the expected position of the magnetic field relative to the Penning trap in Fig. 5.8. By fitting the data to the model to minimise the residuals the final offset was found to be 16 mm, this offset is used in the magnetic field model discussed in Sect. 7.4.1. Alignment of the Babcock magnet relative to the Penning trap is key to capturing in-homogeneity in the magnetic field as well as alignment of the Babcock field to the mirror coils in simulations.

5.2.3 *Axial Mapping of Babcock Magnet Shim Coils*

The next stage of sanity checks looked at the impact of the shimming coils added to the Babcock magnet in 2021. The intention for these coils is to flatten regions of the main coil field symmetrically. To look at the success of this project we repeated the mapping with three different current settings in the shims in Fig. 5.9. Unfortunately, this data set fell victim to testing of internal magnets and has field bumps we attribute to testing in coils E and F with the exception of the red 7 A curve. The general

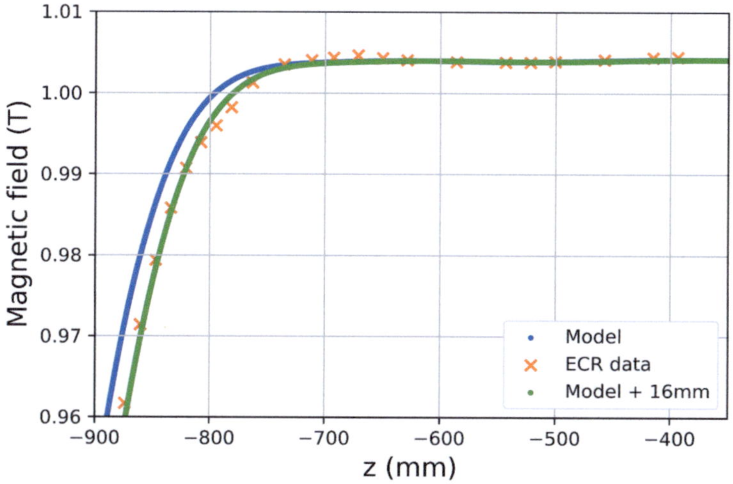

Fig. 5.8 On axis map of the Babcock magnetic field at 191 A with shim coils at 5 A. The ECR data are shown in orange with the original magnetic field model in blue. The green curve shows a 16 mm offset of the position of the Babcock magnet relative to the Penning trap electrodes

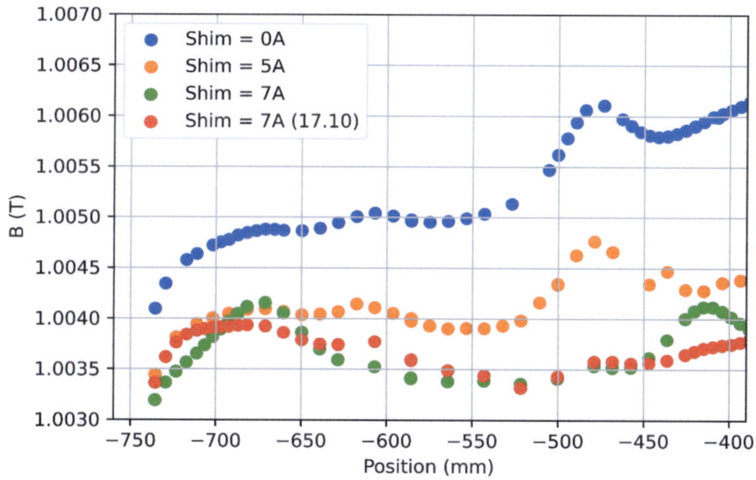

Fig. 5.9 An on axis field map of the Babcock magnetic field at 191 A with varied shim coil currents

trend however shows the flattening of the field as more current is applied. While the increase shim coil current showed promise, due to previously discussed thermal issues in the Babcock (Sect. 3.5.1), the shim coils were set to 5 A despite this not being the optimum field. All data sets from here on will have the 5 A shim field on as a part of the background.

5.2.4 *Axial Mapping of Bottom Short Octupole on Axis Contribution*

While the purpose of the octupole magnets in ALPHA magnetic traps is radial confinement there is a small on axis magnetic field produced by these magnets, particularly at each end where the magnet winding turns. The determination of the on axis contribution of the short octupole field had to take place. This allows a final field value for the gravity ramps to be selected to leave a final magnetic well higher than the largest on axis field from the octupole to avoid reflected antihydrogen changing the z distribution of annihilations. An example of such a final well can be seen in Sect. 7.1.2 and Fig. 7.4. The determination of the octupole field strength was made by taking two axial maps, one with current in the octupole and one without, and taking the difference, the output is shown in Fig. 5.10. Careful attention was made to not have current applied to any of the internal magnets between magnetic field maps to avoid contamination with persistent fields.

5.2.5 *Axial Mapping with High Magnetic Field Gradients*

For the first time in ALPHA, we have the capability of mapping the full magnetic trap. This in part is due to the broad microwave range capabilities as shown in Chap. 4 but also due to the lower axial magnetic trap depth used in ALPHA-g. While the microwave range makes it possible to reach the cyclotron resonance in these fields, this process is not trivial due to the high magnetic field gradients. As the electrons in

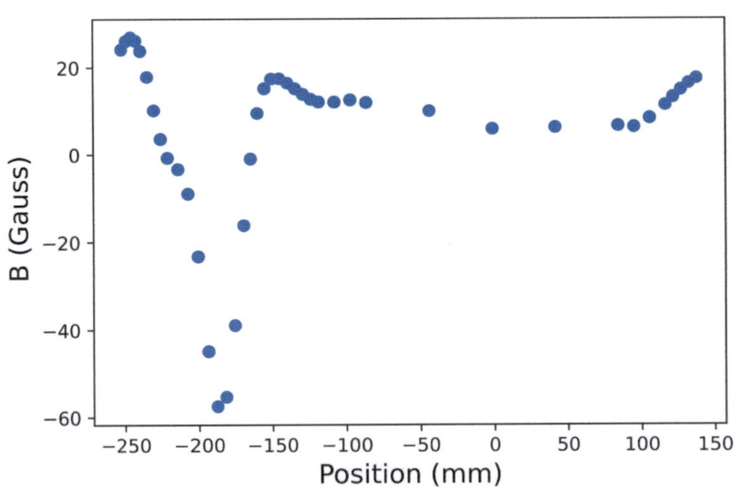

Fig. 5.10 The extracted contribution of the short solenoid along the trap axis axis

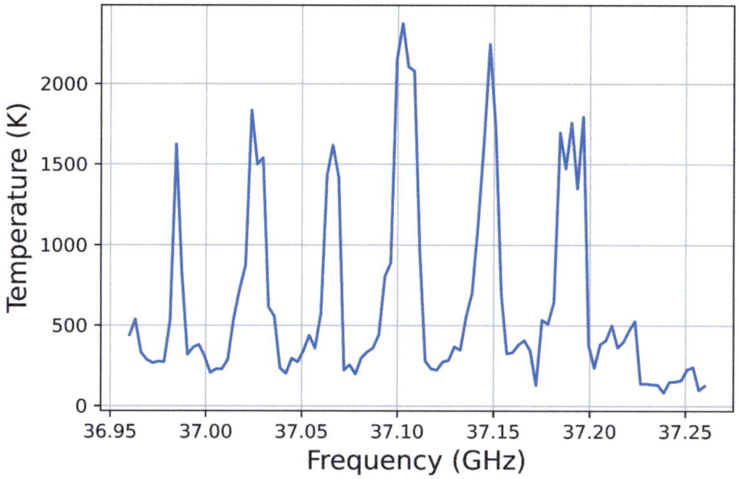

Fig. 5.11 An ECR spectra with 100 points in a high magnetic field gradient showing multiple peaks with no visual identification of the ECR frequency possible

the target plasma have an axial length, when they are confined in an electrostatic well with a high magnetic field gradient the ECR resonance is significantly broadened and many peaks are present. Comparing a low field gradient spectra such as Fig. 5.4 and a high field gradient example in Fig. 5.11 the issue becomes obvious. The peak structures in Fig. 5.11 have no visual indication as to which contains the magnetic field information compared to in Fig. 5.4. The only option in this situation is to adjust the well depth and find the non shifting peak, a time consuming but necessary process.

Along with the shifting bounce peaks it is also essential to the success of these high magnetic field gradient measurements to reduce the axial length of the electron target plasmas. This can be done by several means, first by increasing the size of the applied voltage producing the confining electrostatic potential. Second, is in the design of the Penning trap electrode positions, as discussed in Sect. 4.1.2 the length of all components in the Penning trap electrode assembly were carefully calculated to align the center of electrodes with the center of mirror coils. Finally, reducing the number of electrons per target plasma (by optimising the target extraction potentials as shown in Sect. 5.1.1) reduces the axial extent and therefore the magnetic field the electrons are sampling.

Once the changes to decrease the length of the target plasmas had been made, the first use of this mapping capability is to locate the centre of the mirror coils relative to the Penning trap. This is necessary to allow co-location of the magnets with the detector and therefore determine the "up" and "down" regions which will be discussed in Sect. 6.2.2. The location of the electrodes relative to the mirror coils was constrained to within 0.05 mm of the magnetic model. The location of the electrodes in the detector was determined by confining antiprotons in a deep electrostatic well and detecting annihilations on background gas.

Next, two maps have been made with current in all the magnets used during the release experiments, the currents were chosen so that the magnetic field shape is dominated by the mirror coils. Each map covered the length of the thin trap electrodes shown in Sect. 4.4. The first of these, with mirrors and octupole at full current is shown in Fig. 5.12. There a rough agreement between the measured points in Fig. 5.12 and the expected shape calculated from the initial wire model of the magnets. The measurements in the low gradient regions (near the minima and around the saddle points) are higher precision than the high gradient regions, however the error bars on each point at this scale are unclear. If one examines the difference between the

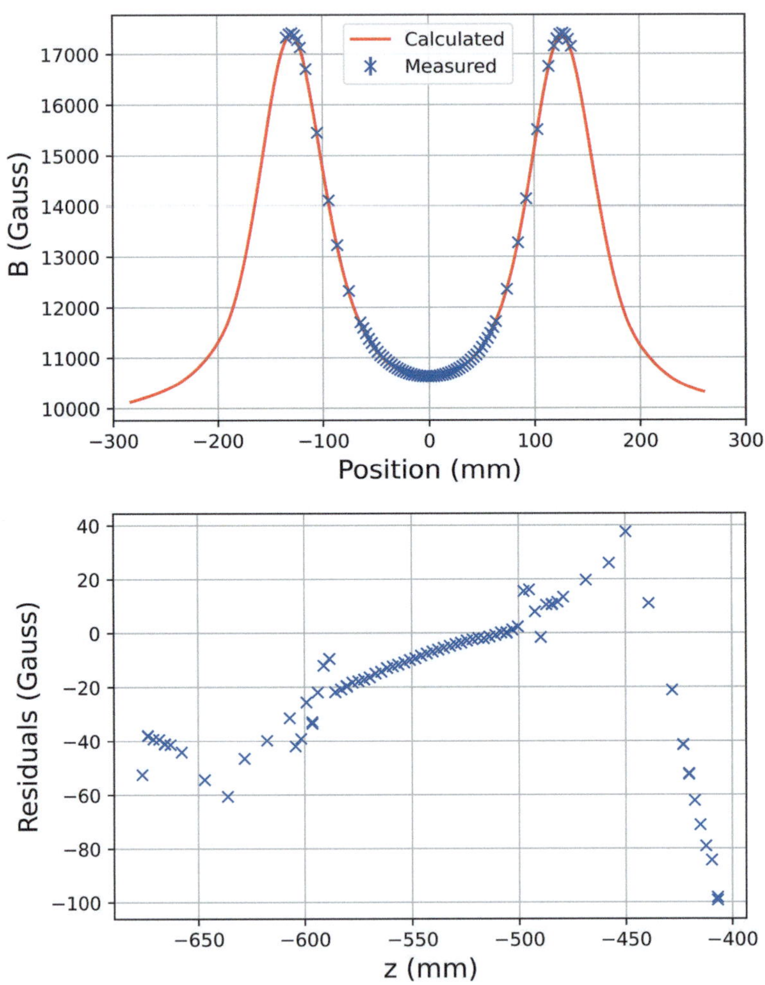

Fig. 5.12 On axis magnetic field map using ECR along the ALPHA-g magnetic trap at full current and residuals from the simulated field before any corrections were applied

calculated field and these data, shown in the figure as residuals, it becomes evident that our understanding of the field is not at the required level of 4 Gauss (clarity on that value will be provided in Sect. 6.1).

A second map, with 10 A in each mirror corresponding to 86% of the magnetic ramp from initial well to final well being complete, was made to to benchmark the magnetic field model. The precision achievable on this map is much greater than that of the full current version as the magnetic field gradients are smaller. However the mirror coil fields are less dominant, meaning errors in models of all other sources (Babcock, shim coils, and short octupole) can be seen here, this will be discussed further in Sect. 7.4.1.

The ability to measure magnetic fields with ECR in a high magnetic field gradient was an essential step for the ALPHA-g experiment and the development of this technique constitutes a major technical step towards the first determination of the free fall of antihydrogen.

5.2.6 Multi-location Simultaneous ECR

Since the entire concept of ALPHA-g relies on precise control of magnetic fields in two locations as a function of time, time variations and delays between magnetic field measurements under each mirror coil can cause a systematic error. The preparation of a reservoir and extraction of target plasmas for a single spectrum can take 2–3 min. If measurements are made under mirrors A and G individually, this leaves several minutes between measurements in the regions of interest that must be characterised carefully. If one can measure at both mirror A and G at the same time, or at least using target plasmas extracted from the same reservoir very close in time, this time difference can be reduced to a few seconds.

In this scheme, two target plasmas are extracted and moved through the trap and to areas of interest, then irradiating with the same microwave pulse (Fig. 5.13). If the magnetic field is the same in each location then the magnetic field information at selected points is recorded at the same time. A difference in measurement time arises when the field is different in each location, i.e. one target is on resonance with the microwaves and the other is not, related to time between irradiation and is limited by the transit time of the targets across the trap. Even with the caveat above this significantly reduced the time (from minutes to a few seconds) between probing fields in different areas.

The first successful demonstration of this technique can be seen in Fig. 5.14. There is strong agreement between measurements made individually and the simultaneous measurements. The magnetic field difference between locations is the difference in the background Babcock solenoid field which must be compensated for during the calculation of the magnetic bias that will be discussed in Sect. 7.1. In theory, this approach can be extended to as many target plasmas as can be held in the Penning trap at any one time. However, this is in practise a challenge due to the range of magnetic fields.

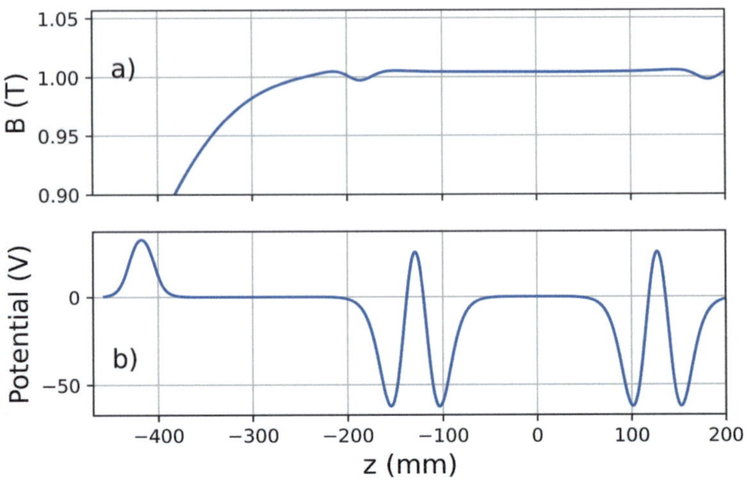

Fig. 5.13 Example of **a** magnetic fields and **b** electrostatic wells used for a simultaneous measurement proof of concept in Fig. 5.14

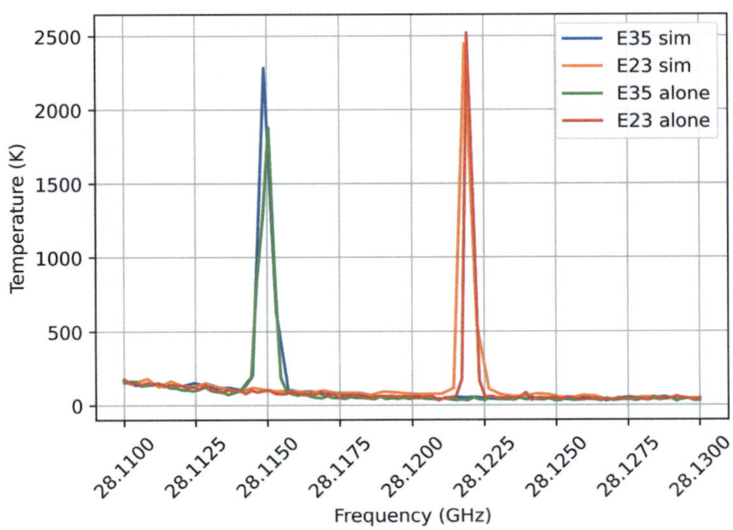

Fig. 5.14 Proof of concept measurement showing the exact agreement between ECR measurements at a single location and simultaneously along the trap at electrodes 23 and 35 as seen in Fig. 4.4

5.3 Summary of ECR Techniques for ALPHA-g

The first ECR measurements in a high magnetic field gradient were essential steps towards a measurement of the magnetic release of antihydrogen. Showing that magnetic fields in this high gradient region can be measured more precisely than the gravitational potential difference across the trap was an initial requirement to be fulfilled by ECR and will be discussed in depth in Chap. 7. The successful implementation of the multi-location ECR set the scene for the ECR measurements to be made during a magnetic release of antihydrogen in ALPHA-g and allow for the characterisation of magnetic field asymmetries, both intentional and uncontrolled. The full set of ECR measurements for ALPHA-g will be summarised in Sect. 7.4.2 following the details of the release experiment itself.

References

1. Amole C et al (2014) In situ electromagnetic field diagnostics with an electron plasma in a Penning–Malmberg trap. New J Phys 16(1):013037. https://doi.org/10.1088/1367-2630/16/1/013037. https://dx.doi.org/10.1088/1367-2630/16/1/013037
2. Hunter ED et al (2020) Electron cyclotron resonance (ECR) magnetometry with a plasma reservoir. Phys Plasmas 27(3):032106. ISSN: 1070-664X. https://doi.org/10.1063/1.5141999. https://pubs.aip.org/aip/pop/article-pdf/doi/10.1063/1.5141999/13806318/032106/_1/_online.pdf
3. Andresen GB et al (2009) Antiproton, positron, and electron imaging with a microchannel plate/phosphor detector. Rev Sci Instr 80(12):123701. ISSN: 0034-6748. https://doi.org/10.1063/1.3266967. https://pubs.aip.org/aip/rsi/article-pdf/doi/10.1063/1.3266967/16070067/123701/_1/_online.pdf.
4. Gabrielse G, Dehmelt H, Kells W (1985) Observation of a relativistic, bistable hysteresis in the cyclotron motion of a single electron. Phys. Rev. Lett. 54:537–539. https://doi.org/10.1103/PhysRevLett.54.537. https://link.aps.org/doi/10.1103/PhysRevLett.54.537
5. Baker CJ et al (2024) Measurements of Penning-Malmberg trap patch potentials and associated performance degradation. Phys. Rev. Res. 6:L012008. https://doi.org/10.1103/PhysRevResearch.6.L012008. https://link.aps.org/doi/10.1103/PhysRevResearch.6.L012008
6. Christensen AJ (2024) Exploiting electron magnetron motion in a Penning-Malmberg trap to measure patch potentials, misalignment, and magnetic fields. PhD thesis, University of California Berkeley. Accessed 16 Sep 2024. https://escholarship.org/uc/item/9v9320jv#author

Chapter 6
The Effect of Gravity on the Release of Trapped Antihydrogen

This chapter will outline the method and results from the first successful experiments in ALPHA-g [1]. I will discuss the concept and methodology for the magnetic release measurement in Sect. 6.1, the development of the experimental procedure in Sect. 6.2.1, the raw data from the first controlled magnetic release of antihydrogen in Sect. 6.2.3, and uncertainties associated with these data in Sect. 6.4. This culminated with the first determination of $a_{\bar{g}}$ as described briefly in Sect. 6.3.

The whole ALPHA collaboration is involved in the operation and data collection for the results in this chapter. My particular contributions relate to operations of the experiment, writing many of the sequences to produce, trap, and release the antihydrogen as well as instigating the protocol for the release measurements. The operation of ALPHA is complex, as such there is an operational structure to ensure smooth running and progress towards physics goals. ALPHA operates with 24-hour shift work performed by researchers, each shift has a leading figure and each week the collaboration appoints an experiment coordinator in control of the physics program for that week and setting work for shift teams. The first experimental "s-curve" was produced during one of my times as experiment coordinator, highlighting my role as a lead scientist in this campaign. I would like to thank my collaborators for this incredible opportunity and experience.

6.1 Magnetic Release of Antihydrogen and the "Balance Point" Concept

ALPHA put the first constraints on the gravitational behaviour of antihydrogen in 2013 [2]. This experiment placed bounds on the possible size of a gravitational interaction to greater than 65 times smaller than matter and less than 100 times greater. These are weak constraints but given the low number of antihydrogen in

A. M. W. Powell, *Magnetic Field Characterisation for Gravitational Free Fall Measurements of Antihydrogen in the ALPHA-g Experiment*, Springer Theses,
https://doi.org/10.1007/978-3-032-07190-3_6

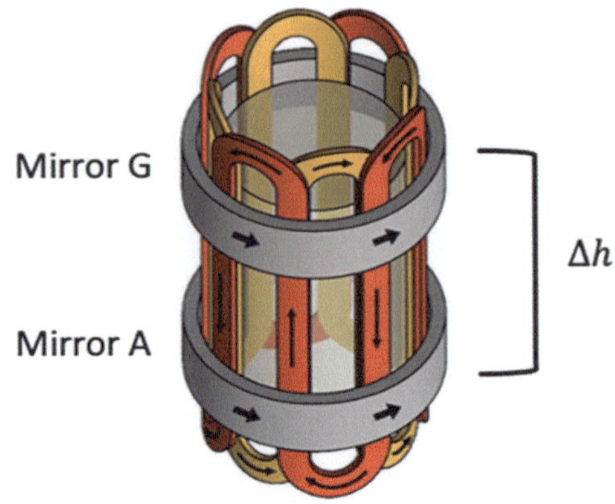

Fig. 6.1 Magnets forming a vertical magnetic minimum trap in ALPHA-g. The winding's of an octupole magnet are shown in orange and gold with two short mirror coils (A & G) in grey

this experiment and the fact that the apparatus was not designed for this style of experiment is not surprising.

The concept of this measurement however led to a new purpose built experiment, aiming to measure this interaction to 1% precision. The proposal for measuring the motion of antihydrogen in Earth's gravitational field in ALPHA-g is using a slow magnetic release from a vertical magnetic minimum trap as shown in Fig. 6.1. As described in Sect. 2.5 the antihydrogen in ALPHA is confined axially using mirror coils. If the magnetic trap is placed vertically there is a gravitational potential between the mirror coils, which are separated by a vertical height Δh, leading to a new overall potential,

$$U = -\vec{\mu} \cdot \vec{B} + m_H g h, \tag{6.1}$$

where m_H is the gravitational mass of (anti)hydrogen, g is the gravitational acceleration. The potential U is sketched in Fig. 6.2a under the assumption that g is the same for matter and antimatter. If one slowly decreases the current, and therefore the confining field, in the mirror coils at the same rate the vertical distribution of annihilations will be biased in the direction of gravitational acceleration as antihydrogen escape. This can be taken further to put limits on the value for the acceleration. If one now repeats this slow magnetic release, but this time with an intentional difference in the magnetic field between the top (mirror G) and bottom (mirror A), there exists a point where the additional magnetic field balances the effect of gravity leading to an equal vertical distribution. In the case shown in Fig. 6.2b this balance point would be where $B_G - B_A = -m_H g \Delta h$. Details of how this is calculated will be provided in Chap. 7. We define a parameter called the bias as,

$$Bias = \frac{\mu_B (B_G - B_A)}{m_H (z_G - z_A)}. \tag{6.2}$$

This is a function of the magnitude of the magnetic field at the maxima of the mirror coil fields, B_A and B_G, and positions of the maxima of the mirror coil fields, z_A and

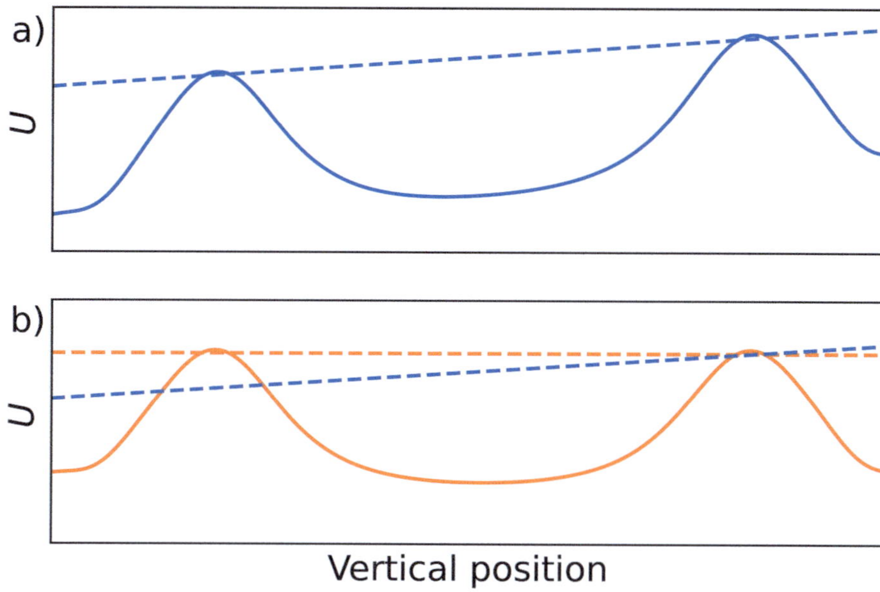

Fig. 6.2 A typical trapping potential, U, for antihydrogen in a vertical trap. **a** Assuming gravity is the same for matter and antimatter **b** no gravity interaction. The dashed blue line in both plots shows a gradient due to "normal" gravity $m_H g \Delta h$

z_G as can be seen in Fig. 6.3. In this one dimensional model, if gravity interacts with hydrogen and antihydrogen in the same manner, one can expect the balance point to be at a value, $Bias = -1$g where g = 9.81 ms^{-2}, throughout this thesis I will refer to the bias in units of g. This intuitive one dimensional model helps to visualise the experiment but does not capture the complexity of the three dimensional magnetic environment. Extensive simulation work is required to account for all effects and map out the balance point, there will be a discussion of how the bias is found using ECR measurements in Sect. 7.1 and how these ECR measurements are used to constrain a magnetic field model for simulations in Sect. 7.4.1.

6.1.1 Defining Up and Down Regions

At this point it is prudent to show Fig. 6.4 visualising the definition of the up and down region, this was determined before release experiments were carried out to avoid bias and are based on the central position of the mirror coils A and G with buffer on either side.

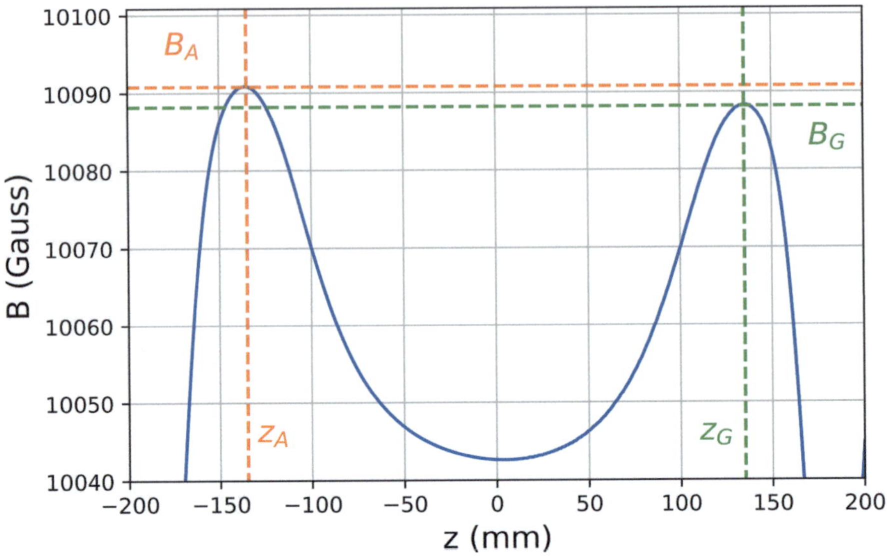

Fig. 6.3 A visualisation of the parameters in the definition for bias used in the magnetic release experimental protocol

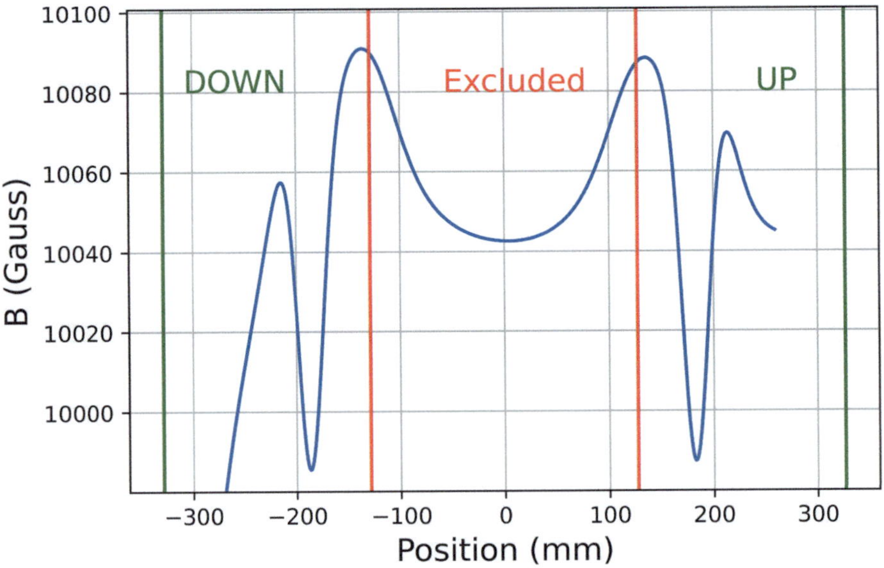

Fig. 6.4 A diagram to show the defined up or down regions overlaid with a typical end of ramp magnetic field profile

6.2 Magnetic Release Experiments

Now that the calibration measurements at high bias value have been made, smaller bias values, between $\pm 10g$ can be used with confidence that the difference in signal between the top and bottom is not related to detector efficiency. This section will lay out the experimental protocol for a magnetic release (using magnet naming convention from Sect. 3.6), show the raw data collected at several bias values, and discuss a second release experiment with a slower magnetic field change.

6.2.1 Experimental Protocol

To ensure a reproducible magnetic environment was available for any release experiments, the following protocol was employed:

1. Program magnetic field profile to select the magnetic bias of the release experiment.
2. Antihydrogen stacking—all magnet currents are applied in a consistent order to form the magnetic trap. Positrons and antiprotons are then slowly merged in an electrostatic potential in the center of the magnetic trap to form antihydrogen, and the process is repeated many times to accumulate a trapped population.
3. Decrease capture solenoid (CapSol) current to 0 A. This magnet is used to increase cyclotron cooling of electrons and positrons during the antihydrogen stacking but produces a field gradient across the magnetic trap that would bias a release experiment.
4. Decrease long octupole (LOc) current to 0 A. This stage has two purposes, firstly it removes a magnetic bias caused by the on axis contribution of the octupole field changing at the end of the magnet windings, for the long octupole this is asymmetric in the bottom magnetic trap as the windings extend across both the top and bottom magnetic trap regions as described in Sect. 3.6. The second reason for this step is to provide a normalisation set of antihydrogen, by removing the LOc, the radial trap depth is reduced and a constant fraction of the trapped antihydrogen population are no longer confined and annihilate. This normalisation will be discussed in Sect. 6.3.
5. Carry out an ECR measurement in initial field before the release experiment. This is detailed in Sect. 7.2.
6. Release antihydrogen by ramping down the top and bottom mirror (MAG) fields while maintaining the magnetic bias. The annihilation position distribution of the antihydrogen in this part of the protocol contains the gravitational information.
7. Carry out an ECR measurement in final field after the release experiment. This is detailed in Sect. 7.2.
8. Decrease short octupole (OcB) and mirrors (MAG) to 0 A. This removes the remaining magnetic trap well and releases any atoms still confined.
9. Measure the background magnetic field with ECR as described in Sect. 7.2.2.

6.2.2 Detector Calibration

To enable the most precise control of the magnet system, the top and bottom mirrors were connected in series following the single mirror release experiments. The mirrors main current will come from a single power supply with a second power supply controlling the bias as discussed in Sect. 3.6. Following this hardware change, a calibration experiment was undertaken, with the current being decreased in both mirrors and a 10g bias applied either up or down. The results of this calibration are used to determine the relative detector efficiency for all future results shown (Table 6.1).

The relative number of annihilation events in each region of Fig. 6.5 during these 10g release measurements is then used to define a relative detector efficiency of up/down as 0.97 (Fig. 6.5).

Table 6.1 Data used for calibration of Up and Down regions of the detector, the magnetic bias was set at −10g and +10g respectively. Fractional counts arise from cosmic background subtraction

Nominal bias	Number of trials	N_{up} (events)	N_{dn} (events)	Loc counts (events)
Up	6	142.9	0.7	169.0
Down	6	−0.1	185.7	213.0

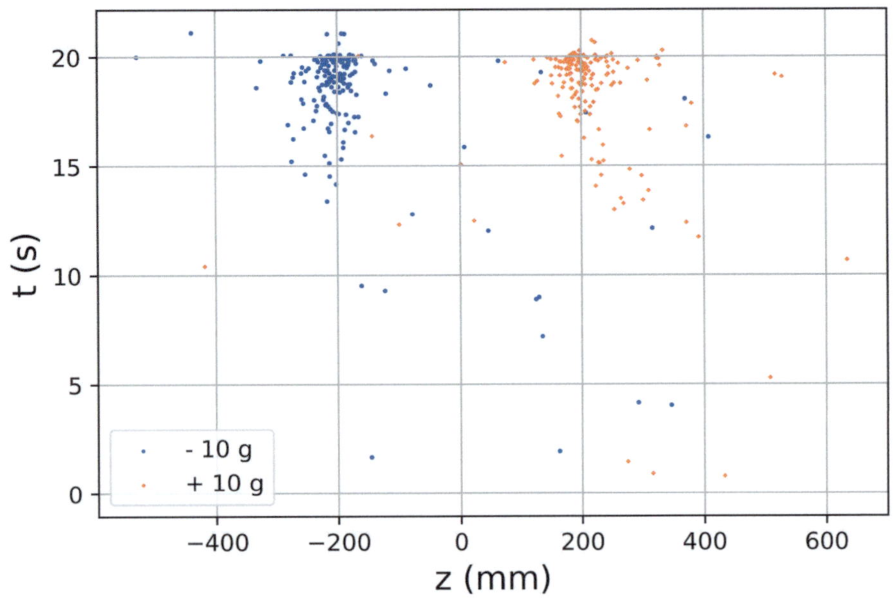

Fig. 6.5 Annihilation events in the detector calibration measurements used to constrain the relative detector efficiency in the Up and Down regions defined in Fig. 6.4

6.2.3 Main Release Experiment

The data used in the determination of $a_{\bar{g}}$ is a set of release experiments with different bias values between $\pm 3g$ with a 20 s magnet ramp time. These data as a function of position are visualised in Fig. 6.6 for bias values between $\pm 3g$ as well as the $\pm 10g$ calibration data. The reconstructed annihilation events in either the up or down regions (as defined in Sect. 6.2.2) and a time cut of 10–20 s during the release is applied and summarised in Table 6.2. There is a background of 0.18 ± 0.01 events and 0.21 ± 0.01 events in the top and bottom regions per trial respectively.

The raw data is converted into a probability of escape downwards, P_{dn}, for each bias value tested as described in detail elsewhere [3]. This probability then forms what we call the escape curve, Fig. 6.7. This escape curve is then compared to simulated escape curves for three different gravitational interactions, normal gravity ($a_{\bar{g}} = g$), no gravity ($a_{\bar{g}} = 0$), and repulsive gravity ($a_{\bar{g}} = -g$). Qualitatively, one can observe good agreement between the normal gravity simulation and these data, while the repulsive gravity simulation is highly excluded.

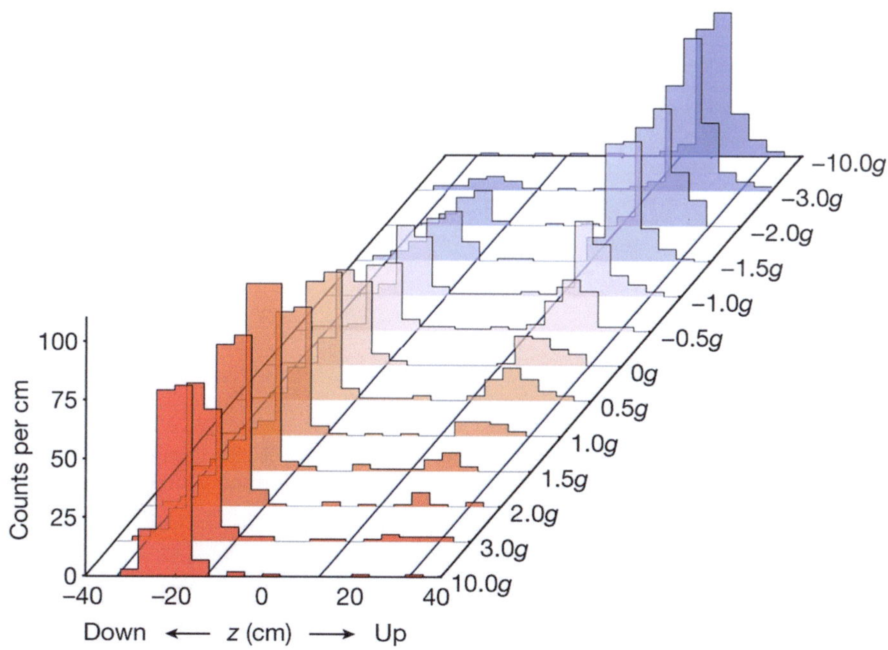

Fig. 6.6 Raw data from the 20 s release experiments showing annihilation event distributions for each bias [1]

Table 6.2 Data for the release experiment trials with adjustment for cosmic background and detector efficiency

Nominal bias (g)	Number of trials	N_{up} (events)	N_{dn} (events)	Loc counts (events)
−3.0	7	151.7	16.5	199.2
−2.0	7	128.7	33.5	168.2
−1.5	6	128.9	57.7	192.0
−1.0	7	69.7	62.5	183.2
−0.5	7	55.7	67.5	201.2
0	7	36.7	94.5	144.2
+0.5	7	36.7	124.5	177.2
+1.0	7	17.7	119.5	185.2
+1.5	6	13.9	180.7	234.0
+2	7	6.7	163.5	228.2
+3	7	7.7	147.5	199.2

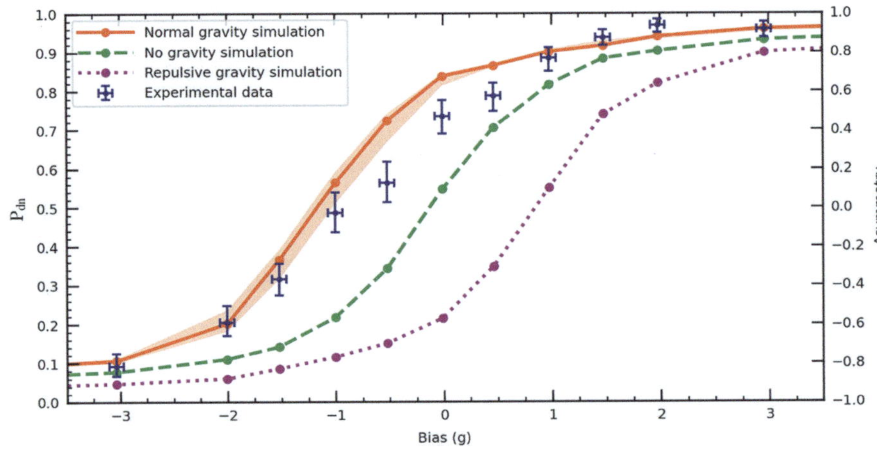

Fig. 6.7 Comparison of the calculated probability of antihydrogen escape downwards for the 20 s release experimental data in Fig. 6.6 to simulations of release experiments with different assumed values of g [1]

6.2.4 *Slow Release Study*

To verify control of the experiment we next increased the release time to 130 s and carried out antihydrogen releases with three bias values. The release curve comparison between the 20 s and 130 s release data points can be seen in Fig. 6.8 along with their relative simulated curves. Due to the time constraints of the antiproton delivery schedule at CERN, only three different bias values could be probed, $0g$, $-1g$ and

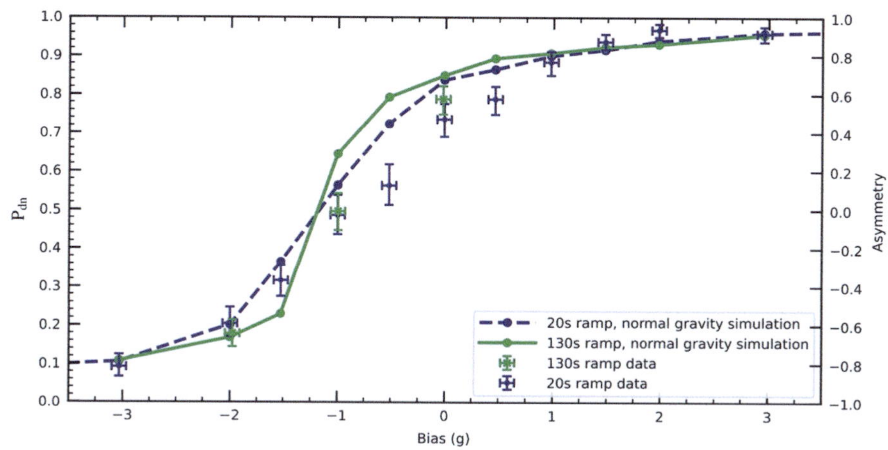

Fig. 6.8 Comparison of 20 s and 130 s release experiments and simulations [1]

$-2g$. These values were chosen based on the 20 s ramp time data set to show the balance point is still constrained between $0g$ and $-2g$ as a systematic check.

Slower release times are ultimately required for a higher precision measurement in ALPHA-g. The longer release time increases the sensitivity of the measurement as the magnetic field changes by smaller amounts per antihydrogen bounce across the trap. Longer release times also lead to increased background during the release window, as the annihilation reconstruction and background rejection is still in early stages of refinement in ALPHA-g, this background quickly becomes too large compared to the number of antihydrogen available. This will improve with increased antihydrogen accumulation and the implementation of the previously mentioned barrel veto by time of flight.

6.3 Determining Value for $a_{\bar{g}}$

I will briefly discuss the determination of a value of the gravitational acceleration of antihydrogen as this will be shown in detail elsewhere. Two analysis teams produced the two methods of extracting a value for gravitational acceleration from the raw data presented. A team from the University of California Berkeley group consisted of Prof. J. Fajans, Prof. J.S. Wurtele, and Dr. D. Hodgkinson. A second team from Brescia University and INFN Pisa was made up of Prof. G. Bonomi, Dr. S. Stracka, and M. Urioni.

A simulation, using the magnetic field model constrained by the ECR results in Chap. 7, is a particle stepping simulation that assumes an energy distribution of antihydrogen on initiation. Particles below the trapping energy survive and are pushed through the magnetic field for a time equivalent to the whole stacking procedure. This ensures that the final energy distribution of particles is representative as energy is mixed between the transverse and longitudinal directions during movement through the field gradients. The particles are then released in the same set of magnet ramps as described for the experimental release for varied values of $a_{\bar{g}}$.

A likelihood analysis of $a_{\bar{g}}$ having a certain value is performed, comparing the experimental data to the simulated escape curves [4]. The value of $a_{\bar{g}}$ was found to be 0.75 ± 0.13(statistical + systematic) $\pm\, 0.16$(simulation)g. The results, within the errors, are consistent with the gravitational interaction of matter and antimatter being equal. It also rules out repulsive gravity or no gravity interaction with antimatter with high statistical significance. An alternative analysis from the Berkeley group, using a modified version of the Bayesian analysis used to find a likelihood of P_{dn} [3], gave consistent results.

6.4 Uncertainties

In the next chapter, I will discuss in depth the uncertainties related to magnetic field measurements and how this translates into uncertainties in the bias value quoted for each data point. In this section, I will discuss the uncertainties in the release experimental data and simulations before showing how these translate to uncertainties in the $a_{\bar{g}}$ value in Table 6.3.

6.4.1 *Systematic and Statistical*

There are two main sources of statistical and systematic errors identified: the finite data size in the release experiments and the calibration of the detector.

Finite data size: due to the trapping rate and antihydrogen production in ALPHA-g. Approximately half the antihydrogen per trial were also released during the LOc ramp down and acted as a calibration for total particle number in a trial. The error is assigned as a square root of the number of counts.

Relative detector efficiency: the difference between the top and bottom region detection efficiency is contained in the calibration error. This difference is due to the calibration data set having a low number of antihydrogen to constrain the relative efficiency well as can be seen in Table 6.1.

These uncertainties are shown as vertical error bars on P_{dn}in Fig. 6.7.

6.4.2 *Simulation*

As the trapping potentials are three dimensional, the simple one dimensional model discussed in Sect. 6.1 is not sufficient to describe the escape point of antihydrogen

Table 6.3 Summary of uncertainties in the value of $a_{\bar{g}}$ shown in Sect. 6.3. They are constrained by statistical, systematic, and simulation errors described in Sect. 6.4

Uncertainty type	Uncertainty description	Magnitude (g)
Statistical	Finite data size	0.06
Statistical and Systematic	Calibration of detector efficiencies in up and down regions	0.12
Systematic	Other minor sources	0.01
Simulation	Modelling of magnetic fields	0.16
Simulation	Antihydrogen initial energy distribution	0.03

atoms. The motion of antihydrogen in the trapping potential during a release experiment is simulated and a data driven adjustment to the magnetic field profile used in the simulations was made. Details of that adjustment are in Sect. 7.4.1. The error constraints in this simulation are shown as an orange band on Fig. 6.7 and arise from:

Modelling of magnetic fields: adding perturbations to the magnetic field model that adjust the off axis magnetic field, while remaining consistent with the data used to adjust the on axis field model. These include: multipolar perturbations, off-axis persistent fields, and transverse offsets of mirror coils (MAG) relative to the short octupole (OctB).

Initial energy distribution of the simulated antihydrogen: this is experimentally unmeasured as the process of confinement of the formed antihydrogen during mixing is not fully understood. Limits on the size of this error were found by bootstrapped simulations with subsets of the original population to show the maximum effect of an incorrect distribution.

All sources of errors discussed in this section and in Sect. 7.4.2 contributed to the determination of $a_{\bar{g}}$ and allow for estimated uncertainties in that value summarised in Table 6.3.

6.5 Events Outside the Release Window

As the annihilation reconstruction and background rejection improved it became apparent that there were annihilation events outside of the magnetic release windows. Most notably for this thesis, some of these events appeared during the ECR measurements before and after the mirror ramp as will be shown in Sect. 7.2. The example shown in Fig. 6.9 is for the ECR measurement window after the mirror A and G ramp summed for all $\pm 2g$ bias release experiments. An initial suspect in the cause of these events was the microwave pulses used in ECR were driving a positron spin flip in the antihydrogen, making it high field seeking and therefore become ejected from the trap. The first indication that this is not microwave related can be seen in the figure as most of the counts are before microwave irradiation begins. Secondly, if one calculates the frequency required for these transitions for the magnetic fields being

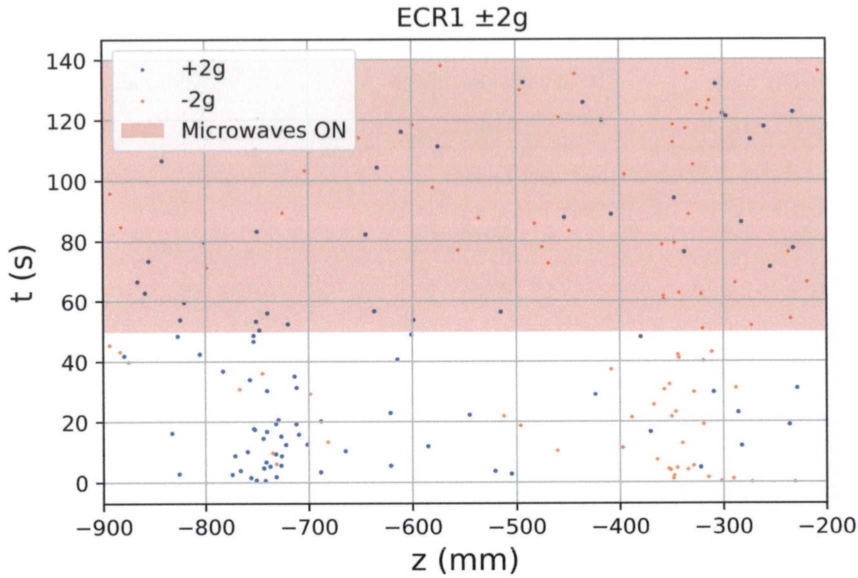

Fig. 6.9 Annihilation events following magnetic release experiments at $\pm 2g$ with a highlighted region showing the time of microwave irradiation for ECR measurements

probed it becomes obvious there is no overlap due to how shallow the magnetic trap is following the end solenoid ramps. There are several working theories around the cause of these losses and extensive simulations have been completed in an attempt to reproduce the effect. If the addition counts are compared to the escape curve, as in Fig. 6.10, the overall position of the counts follows the bias values in the same way as the escape curve.

6.6 Summary

The results in this chapter mark a significant breakthrough with the first direct observation of the effect of gravity on antimatter with ratio of $a_{\bar{g}}/g = 0.75 \pm 0.13$(statistical + systematic) $\pm\, 0.16$(simulation) being extracted from the release trials. This result is consistent with hydrogen and antihydrogen having the same gravitational acceleration within errors. These constraints have allowed the ALPHA collaboration to rule out two possible extreme differences between the interaction of matter and antimatter with gravity; no gravity effecting antimatter and repulsive gravity have both been ruled out with high significance.

This measurement in fact surpassed the initial aims of the collaboration to determine the sign of the gravitational acceleration in a so called "Up vs Down" experiment.

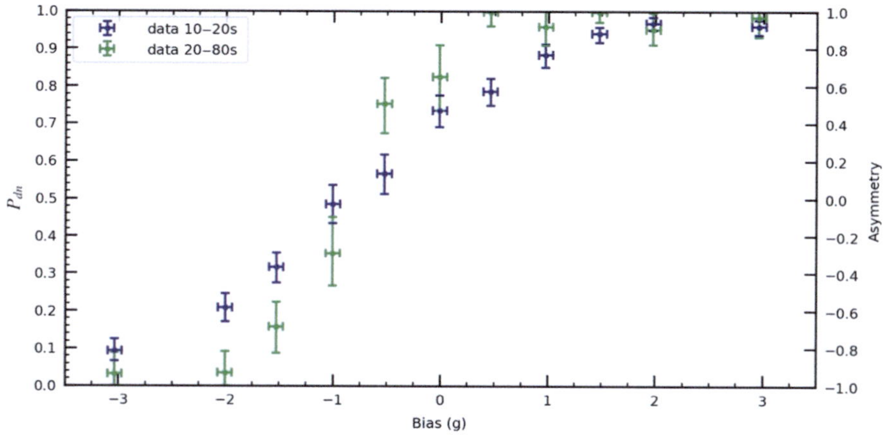

Fig. 6.10 Visualisation of annihilation events outside the release window as a function of the bias [1]

The work in Sect. 6.2.4 is a starting point for systematic studies in these release experiments. The data presented in this section shows the response of the s-curve shape to the release time is consistent with simulations and offers a first path to a higher precision measurement.

References

1. Anderson EK et al (2023) Observation of the effect of gravity on the motion of antimatter. Nature 621:716–722. ISSN: 1476-4687. https://doi.org/10.1038/s41586-023-06527-1
2. Charman AE (2013) Description and first application of a new technique to measure the gravitational mass of antihydrogen, Nat Commun 4(1785):1–9. ISSN: 2041-1723. https://doi.org/10.1038/ncomms2787
3. Hodgkinson D, Fajans J, Wurtele JS (2023) A Bayesian data analysis method for an experiment to measure the gravitational acceleration of antihydrogen. Phys Sci Forum 9(1). ISSN: 2673-9984. https://doi.org/10.3390/psf2023009009. https://www.mdpi.com/2673-9984/9/1/9
4. Urioni M (2024) Measurement of Earth's gravitational acceleration on anti-hydrogen with the ALPHA experiment at CERN. PhD thesis. Università Degli Studi di Brescia

Chapter 7
ECR for Gravity Experiments

The preceding chapters have laid the foundations for the results that will be shown in this chapter. The ECR magnetometry technique was outlined in detail in Chap. 5 and the gravitational experiment results published by ALPHA in 2023 [1] and shown in Chap. 6. This chapter pertains to all ECR data used in the characterisation of the magnetic field in those gravitational experiments, namely in setting the magnetic bias at the start and end of a magnetic release plus systematic studies to constrain what the bias looks like between these points. Section 7.1 will discuss selection of magnet currents to produce the desired magnetic field biases, Sect. 7.2 shows the magnetic field measurements made during the release experiments, Sect. 7.3 has details of ECR measurements made to constrain uncertainties along with the techniques developed to acquire those data. Section 7.4 will contain a discussion of how these data were used to adjust the magnetic field model used in simulations of the antihydrogen trajectories in ALPHA-g along with a summary of the final uncertainty values. Finally Sect. 7.5 contains details and results from proof-of-principle ECR measurements designed to inform future higher precision experiments in ALPHA-g.

The data contained in this chapter was collected by several members of the ALPHA collaboration with my instruction and supervision. I developed the experimental procedures for the data collection with contributions from A. Christensen and Dr. Chris Ørum Rasmussen. Dr. Chris Ørum Rasmussen and Dr. C. So were heavily involved in the quantification of uncertainties.

A. M. W. Powell, *Magnetic Field Characterisation for Gravitational Free Fall Measurements of Antihydrogen in the ALPHA-g Experiment*, Springer Theses,
https://doi.org/10.1007/978-3-032-07190-3_7

7.1 Adjustments to Magnet Currents for Varied Magnetic Bias

The first role for ECR in the gravity measurement campaign itself was to find the mirror coil currents required to produce the variety of bias values described in Sect. 6.1. All measurements were carried out with simultaneous ECR under the top and bottom mirror coils (see Sect. 5.2.6) to minimise any errors from time dependent fields. The response of the field in both locations from applying current in the bias power supply (Sect. 3.6) is tracked in Figs. 7.3 and 7.1. As expected there is linear response of the bias to the current, the raw data shows that the field under the top mirror coil changes linearly with negligible shifts in the bottom mirror coil field.

7.1.1 *Initial Bias*

The starting point of the ramp is from the initial magnetic well. The magnetic field in the initial well is from the mirrors A and G at full current (70 A) then adding or subtracting field with the bias supply (as described in Sect. 3.6) on mirror G. The modelled magnetic fields for each bias in the initial well can be seen in Fig. 7.2. The current required in the bias supply to achieve each bias value is shown in Fig. 7.3.

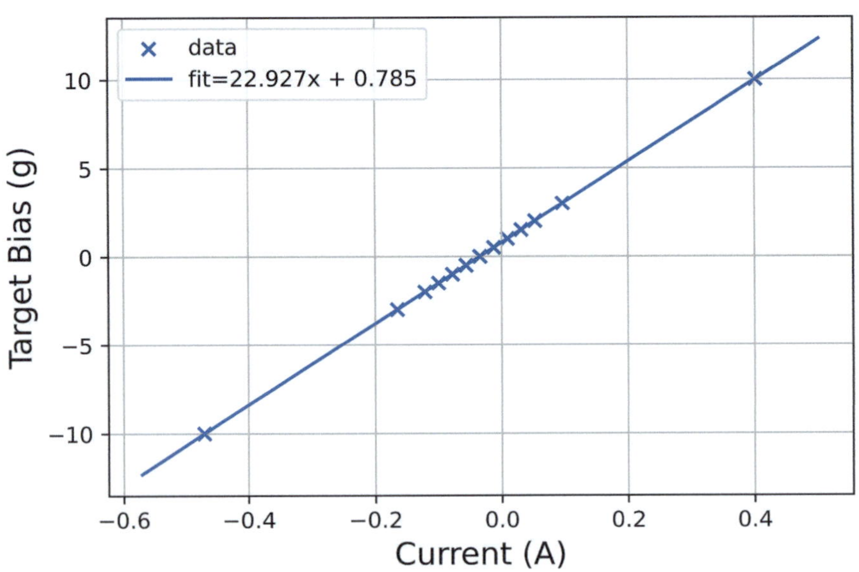

Fig. 7.1 Control of the magnetic bias as a function of current applied to the mirror G bias supply for the final magnetic field after a release

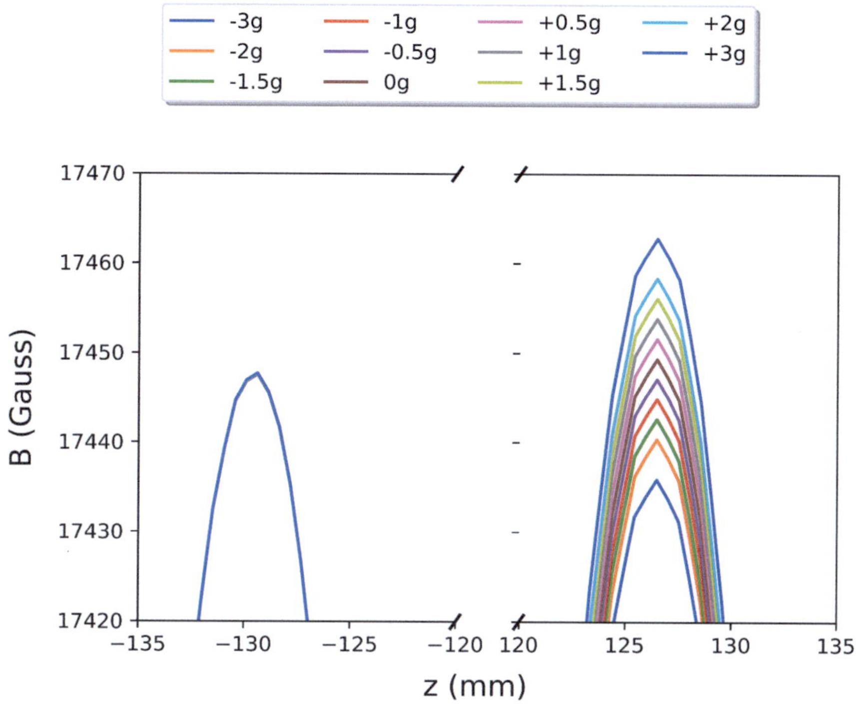

Fig. 7.2 Calculated magnetic field profile showing the maxima formed by mirror A and mirror G with different currents in the mirror G bias supply forming the range of bias values

7.1.2 Final Bias

Finding the magnet current configurations for the bias in the final well is a slightly more involved process, since the magnetic field provided by mirrors A and G is reduced and shifted during a release due the background axial field created by the short octupole and external solenoid. The axial field from the short octupole begins to dominate the field under the mirrors and move the axial position of the saddle point and making the change in magnetic field non-linear. An example of this effect can be seen in Fig. 7.4 which shows the modelled on axis field at the end point of a ramp. In order to minimise the non-linear section of the ramp down due to the on axis contribution of the octupole field, the final well is not at 0 A in the mirror coils. The difference between the bottom of the well (orange) and the top of the lower mirror (green) field is set to be large enough such that any antihydrogen that have escaped during the release aren't confined by the on axis contribution from the octupole end turns and can escape fully. The bias in the final well was found by making adjustments to the current supplied to mirror A and G in series and to the bias supply on mirror G. Field models for the final well bias' can be seen in Fig. 7.5.

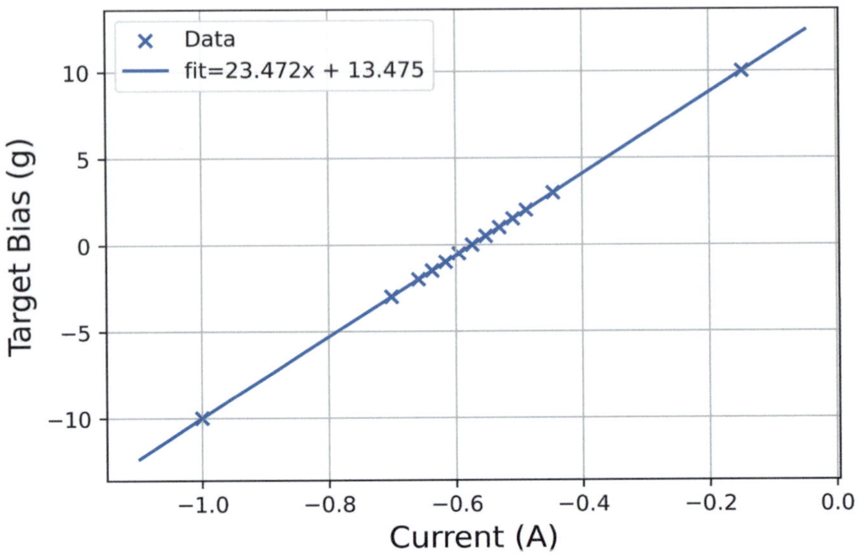

Fig. 7.3 Control of the magnetic bias as a function of current applied to the mirror G bias supply for the starting magnetic field

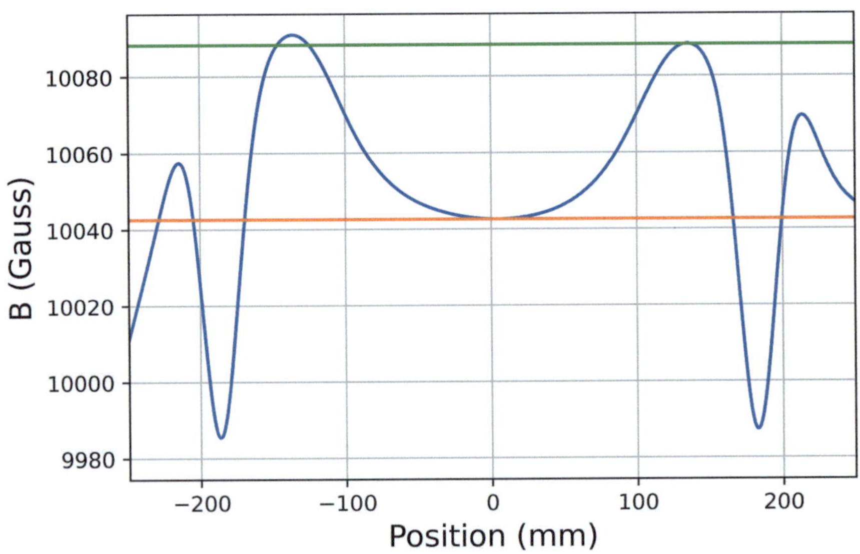

Fig. 7.4 Visualisation of the final magnetic field profile during a typical release experiment. The difference between the green and Orange lines is labeled as the final well depth and is between 40 and 50 Gauss

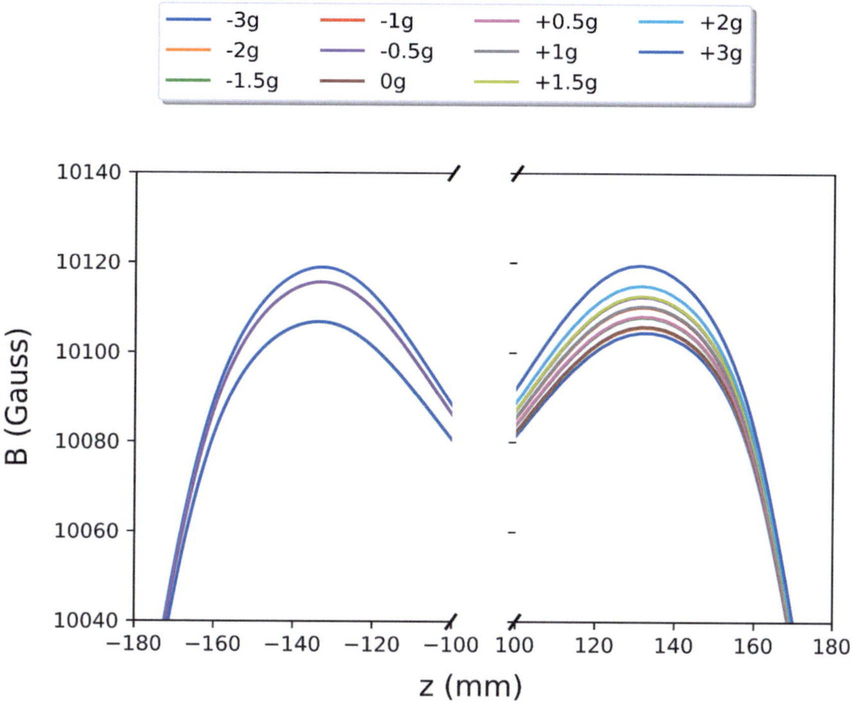

Fig. 7.5 Calculated magnetic field shapes produced at the different bias values

The current required in the bias power supply to achieve each value of the bias can be seen in Fig. 7.1.

As the length of the neutral trap is changing as a function of magnetic field, this raises an issue that the definition of the difference in gravitational potential between the release points is changing during the experiment and that the magnetic bias itself changes during the release. This systematic effect has been studied through axial field maps under both mirror coils at every bias value (Sect. 7.3.1) and summarised in the final uncertainties Table 7.1.

Table 7.1 Summary of magnetic field uncertainties from ECR, magnetronmetry and field modelling

Uncertainty	Magnitude (g)
ECR spectrum width	0.07
Repeatability of ($B_G - B_A$)	0.014
Peak field size and z-location fit	0.009
Field decay asymmetry (A to G) after ramp	0.02
Bias variation in time	0.02
Field modelling	0.05

7.2 ECR as a Field Diagnostic During a Release Experiment

During a single trial, or release, for the data to be presented in the escape curve three ECR measurements were made. The first was taken at high field, following the ramp down of the current in the booster solenoid and the long octupole magnet. The second was taken immediately following the end of the ramp of the top and bottom mirror coils. Each of these measurements were performed under mirrors A and G using the simultaneous technique discussed in Sect. 5.2.6. An example of the magnetic field and electrical potentials used during these measurements can be seen in Fig. 7.6, summary results per bias are shown in Figs. 7.7 and 7.8.

7.2.1 *Initial Field Versus Final Field Tracking Measurements*

Initial well measurements made during the release protocol (Sect. 6.2.1) are lower precision than most ECR measurements made in ALPHA. The larger FWHM of the fit initial field measurements can be seen easily in Fig. 7.9. There are several contributions to the broadening of the high field measurements, firstly the field gradient the electrons are sampling in the high field broadens the resonance. Secondly, due to the previously discussed microwave standing wave pattern issues means that to

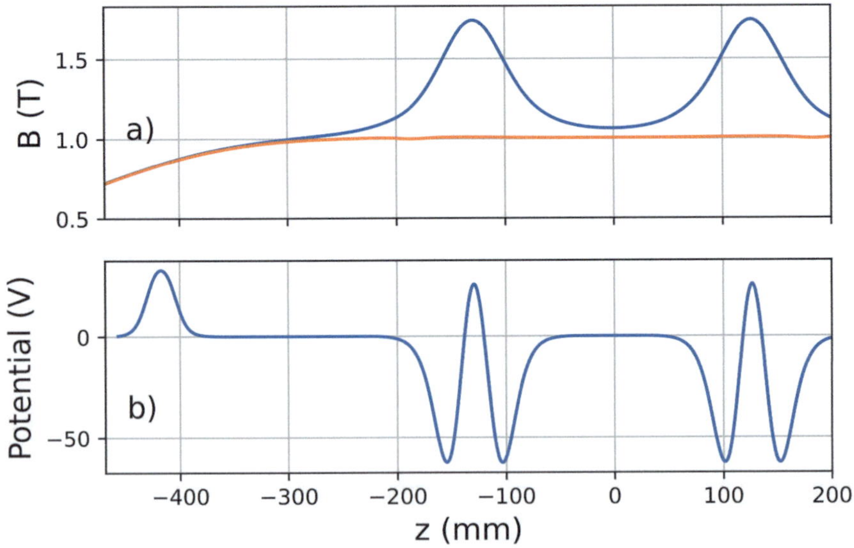

Fig. 7.6 **a** On axis magnetic field profile at the start (blue) and end (orange) of a release experiment. **b** On axis potential used to confine the reservoir and target plasmas at each irradiation step of the ECR measurements during a release experiment

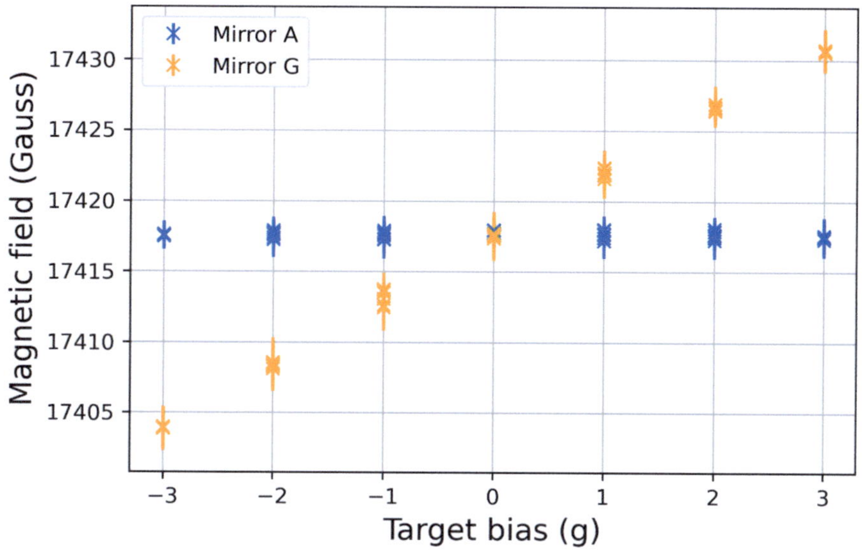

Fig. 7.7 Results of ECR field measurements in the initial magnetic field (before release) for all bias values

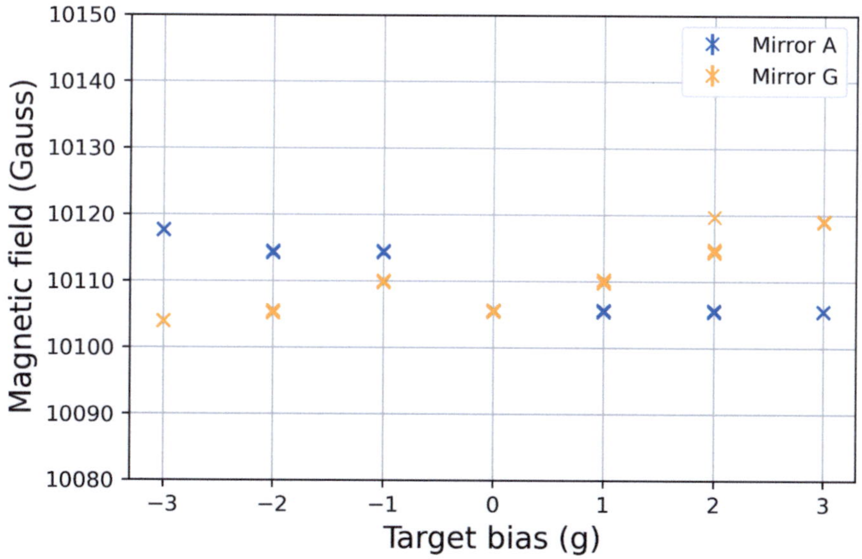

Fig. 7.8 Results of ECR field measurements in the final magnetic field (after release) for all bias values

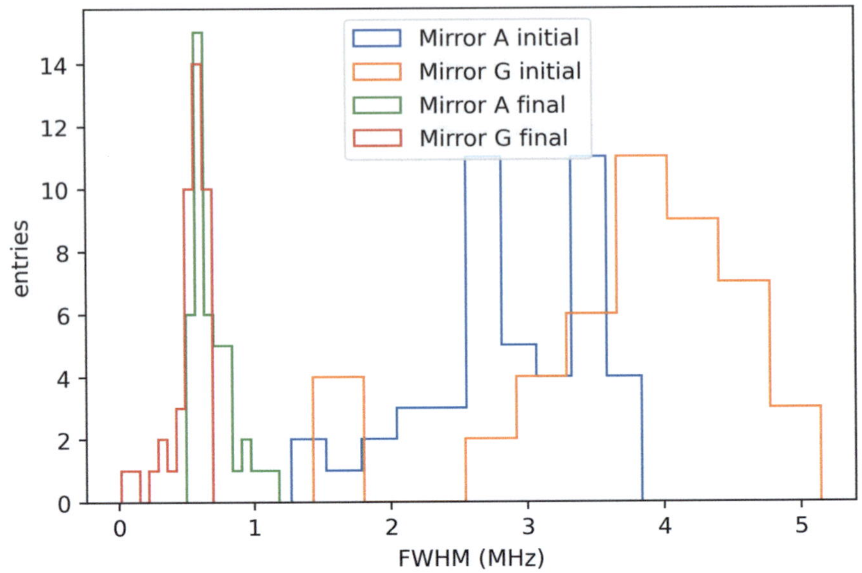

Fig. 7.9 The full width at half maximum (FWHM) of all ECR measurements at the initial (high) and final (low) magnetic fields during the main release experiment

ensure adequate heating at both locations at the same time some measurements have a much higher microwave power than required, again broadening resonance.

7.2.2 *Background Field Tracking*

The third measurement, a background tracking of sorts, was made after all magnets had been set to 0 A and acted as a marker to look for changes in the overall field over time. The background measurements over time show a mostly consistent difference between the fields throughout all experimental runs and can be seen in Fig. 7.10. There is an as of yet unexplained magnetic anomaly that seemed to decay over many days. The working theory is that this is related to a magnet ramp that was aborted part way through causing a fast magnetic field change and creating some persistent currents in the Babcock magnet. Extensive testing of this hypothesis has not been carried out yet but since the decaying field was symmetric under the top and bottom mirror coils this did not influence the data set.

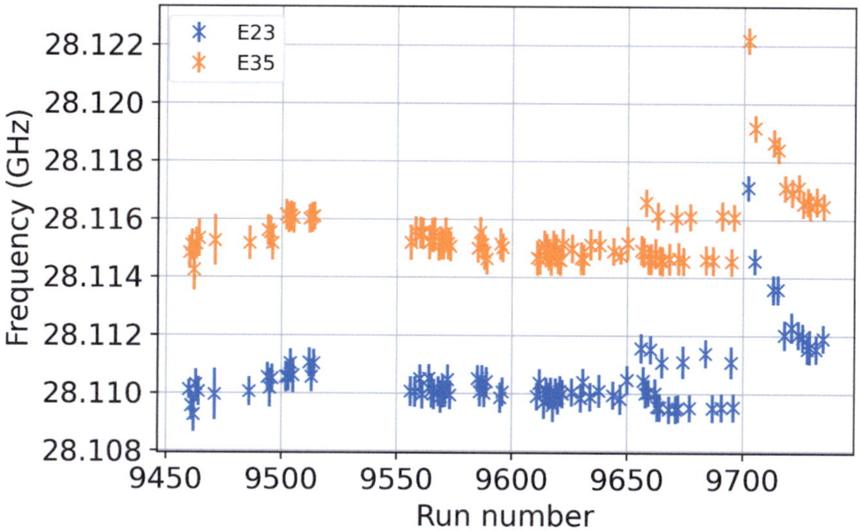

Fig. 7.10 ECR measurements of the background field after all release experimental trials

7.3 Studies for Constraining Bias Uncertainties

7.3.1 *Saddle Point Mapping*

This section will show results from the mapping around the maxima of mirror coils A and G in the final well for various bias values. This adds a constraint on the change of trap length discussed in Sect. 7.1.2. The method was to ramp the magnets to high current, as would occur during a release experiment, then down to the final well with a set bias value. The magnetic fields around the saddle points of mirrors A and G were mapped using ECR. An example axial map around the designed center of each mirror coil can be seen in Fig. 7.11. A point at the designed center of the magnet/electrode was repeated multiple times during the map to track the decaying magnetic field, which was then subtracted from the measured points in each map. Figure 7.12 shows the changing positions of the magnetic maximum produced by either mirror A or G for each bias. From this a trap length can be associated with each bias and the actual bias at the end of the release is extracted from the fits.

The minimum of the final well was also mapped out to find the final well depth, allowing for comparison to the magnet model and comparison of simulation of remaining antihydrogen atom in the final well to experimental observations. These measurements were repeated for all bias values used in release experiments.

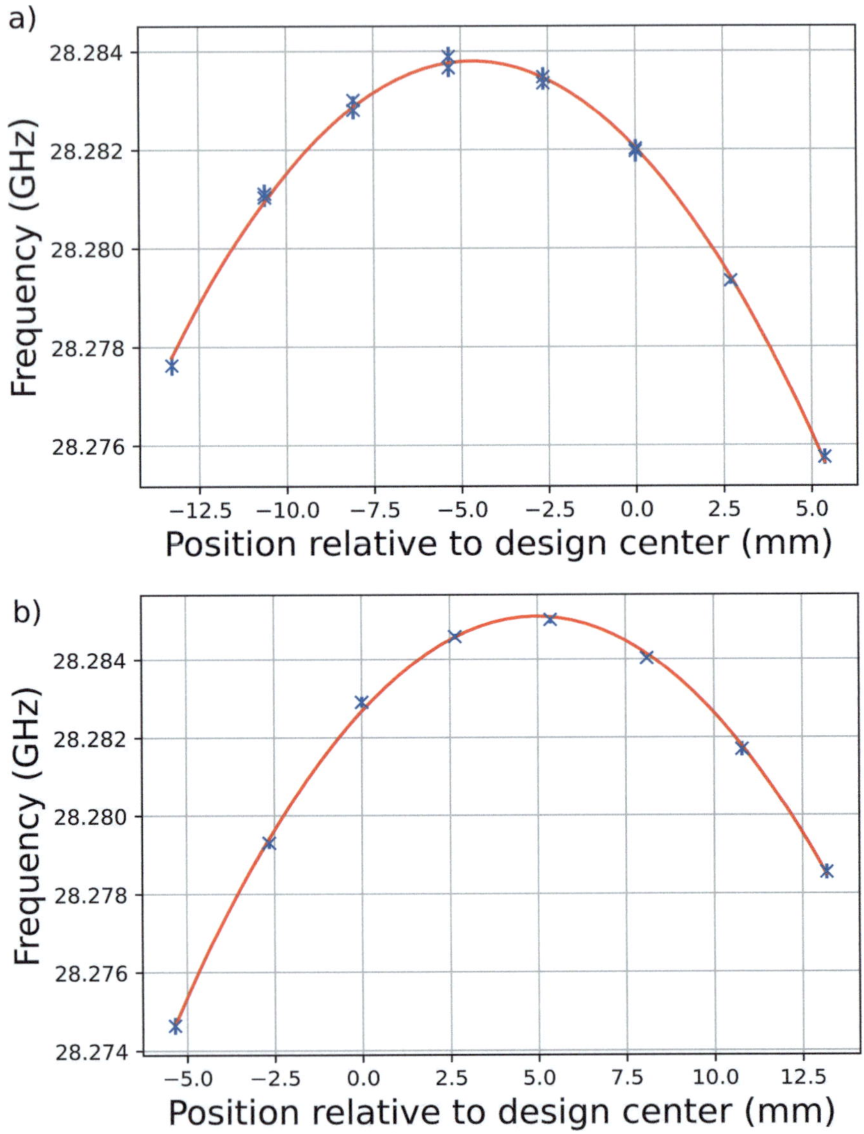

Fig. 7.11 Example axial maps of **a** mirror A and **b** mirror G at the end of a magnetic release for a 0g bias

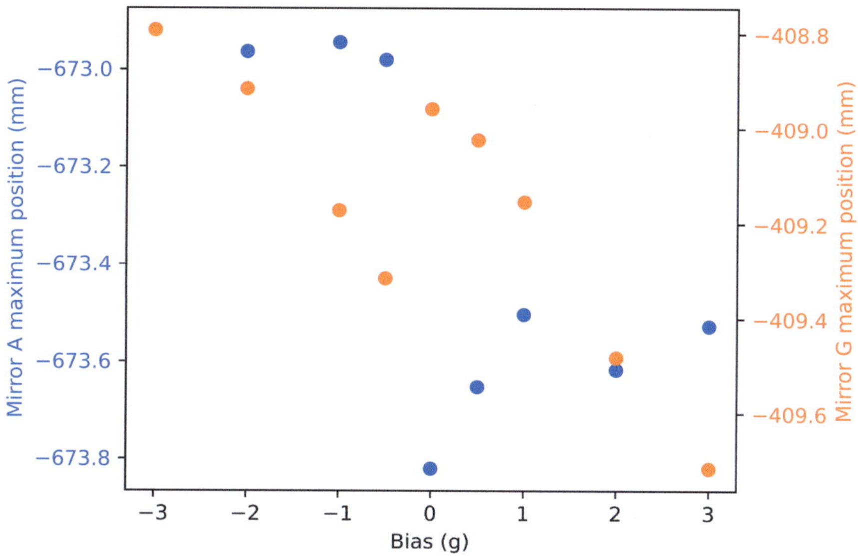

Fig. 7.12 Positions of the maximum produced by mirror A and G at the end of the magnetic release for all bias values

7.3.2 Fast Repeat ECR

As previously discussed a major issue faced in the design of ALPHA-g was minimising the persistent fields that would be induced during a release. While these persistent fields can be minimised they cannot be completely eliminated and must be characterised through measurements. Most importantly, there must be constraints placed on any asymmetries between magnetic fields under mirrors A and G. To constrain the difference in these persistent field I have developed a modified ECR we are referring to as "fast repeat ECR". To reduce the time between field measurements I aimed to make multiple measurements per reservoir, removing the time taken to load and prepare the electron reservoir.

In a proof of principle measurement I used 50 target plasmas per field measurement and 200 target plasmas per reservoir. This translates to four repeats of the same frequency sweep and therefore four measurements with each reservoir separated by around 20–30 s. The reservoir was loaded before the field ramp was initiated such that the first microwave pulse was sent to the trap less than 500 ms after the ramp had ended. A set of sweeps can be seen in Fig. 7.13. This example is immediately following the end of the magnet ramp and one can see an obvious decrease in the field during each sweep. The relative peak heights seen are related to the changing number of electrons in the target plasmas. The reservoir was optimised to start the target plasma production quickly after electron loading was complete but in this case the reservoir waits for the 20 s magnet ramp, causing the initial target plasmas to be small.

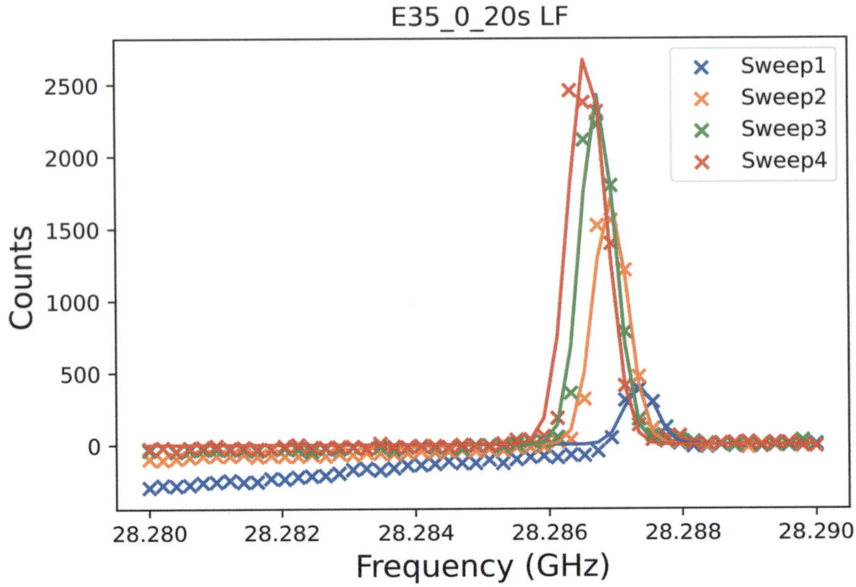

Fig. 7.13 Proof of principle fast ECR data

A refined version of the measurement was carried out for three bias values. In this second version, the microwave frequency sweep was set from high to low, thus as the field decays the heating appears in later scoop numbers. It is preferable to have the first data point as close to the end of the magnet ramp as possible to best constrain the initial fast decay. In this improved measurement technique, eight microwave sweeps per reservoir were carried out, more data points per reservoir again helps to constrain the decay curve. An example of the measurement output for a single reservoir in the fastest part of the decay can be see for a 0g bias run in Fig. 7.14. The final combined data set used to extract an error is in Fig. 7.15. The ΔB is the magnetic field decay asymmetry and this error will be discussed in the summary Sect. 7.4.2.

7.3.3 Stepped ECR

To ensure the induced currents during the ramp did not drastically impact the field as it changed, we measured the ECR frequency during stepped ramp up or ramp downs. Care was taken to ensure a magnet ramp history consistent with the magnetic release experiments prior to the interruption of the ramp down, ensuring the same persistent fields were present.

The magnet ramp from the initial well to the final well was interrupted at 10 points and ECR measurements made at both mirror A and G. This was repeated for five different bias values and averaged to provide an estimate of the error for all values

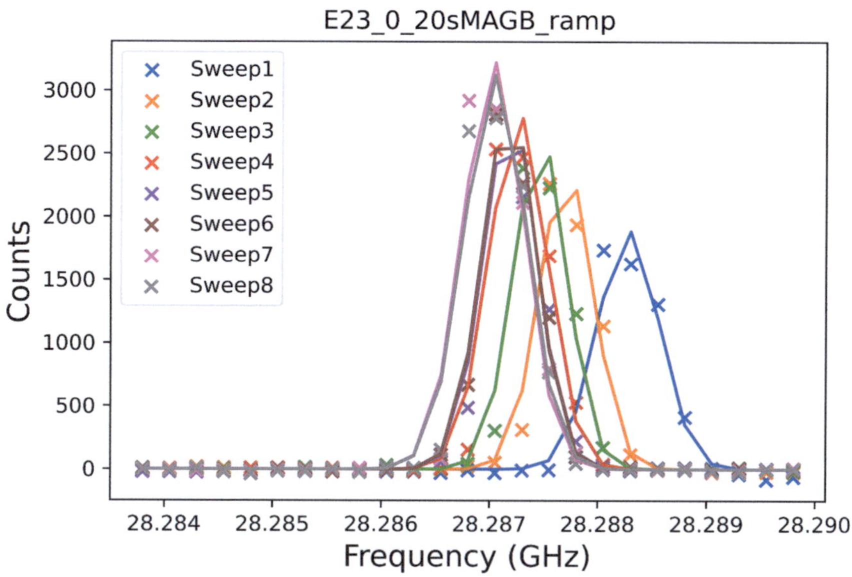

Fig. 7.14 Refined fast ECR measurement

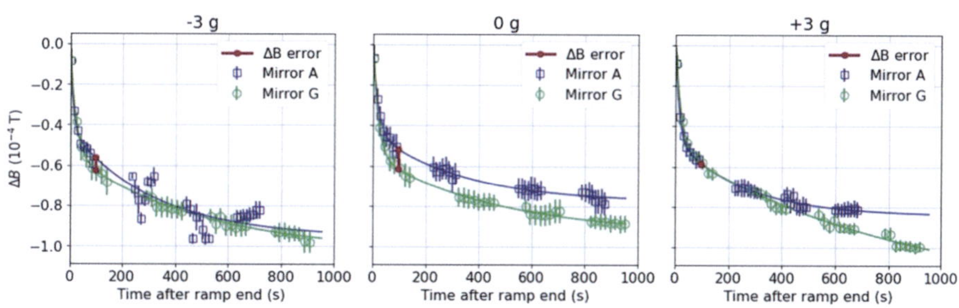

Fig. 7.15 Final fast repeat ECR used to constrain difference in magnetic field decays under the top and bottom mirror coils

of the bias. A caveat to these measurements is that no drift from decaying fields was quantified and subtracted as in other measurements for this chapter. This allows for an error in the measurement from time varying fields that changes depending on the point of the ramp that is measured.

Despite this caveat, data collected shows that there is a deviation from linear during the ramp down, consistent with a build up of induced currents. These data combined with a new method of measuring magnetic fields using the magnetron motion of target plasmas by Andrew Christensen are used to constrain the size of the deviation and integrate it into the magnetic field model.

7.4 Field Modelling and Summary of Uncertainties in Bias

7.4.1 Data Driven Magnetic Field Model Adjustments

Given all the data gathered during this campaign, an intensive effort was made to include all known effects in the magnetic field model. This model, created by Dr. Chukman So, is based on the construction of the magnet windings and is used in the simulations of the antihydrogen trajectories during a release. Images of each layer of the magnet windings were taken during manufacturing and from these images accurate dimensions were taken. Based on the measurements presented in Sect. 5.2.2, the position of the Babcock relative to the BNL magnets was adjusted by 16 mm and a final field B_{Babcock} was determined which includes the contributions from the main solenoid at 191 A and the shim coils at 5 A. The term B_{SOct} is the model for the short octupole magnet, the model has been adjusted to fit the data shown in Sect. 5.2.4. The position of mirrors A and G relative to the electrodes showed an agreement with the model to within 0.05 mm as in Sect. 5.2.5. The field models for each mirror are labelled B_{MAB} and B_{MGB} and are optimised to fit the ECR data in Sect. 5.2.5 where the mirror currents are dominant. I_{MAB} and I_{MGB} are currents applied to each mirror and allow for small current adjustments to fit the model to data with different mirror coil currents set.

The wire model then has two additional terms related to the uncontrolled magnetic fields as shown in Eq. 7.1,

$$\begin{aligned} B(z, p) = & \overbrace{B_{\text{Babcock}}(z) + B_{SOct}(z) + B_{MAB}(z)I_{MAB}(p) + B_{MGB}(z)I_{MGB}(p)}^{\text{Wire model}} \\ & + \underbrace{A(z)(1 - exp(-p/0.1346))}_{\text{Magnetronmetry term}} + \underbrace{B_{Res0}(z)(1 - p) + B_{Res1}(z)p}_{\text{Residual to account for ECR}}. \end{aligned} \tag{7.1}$$

The first term is related to the build up and saturation of persistent fields during a ramp as characterised by the new “magnetronmetry” technique developed by A. Christiensen. p parameterizes the fraction of the ramp completed, from 0 (initial well) to 1 (final well) and A is a term found experimentally. I will not cover the details here as this will be discussed elsewhere and just summarise by saying the technique uses the same electron target plasmas as ECR and tracks the magnetron motion to extract a magnetic field. The targets are shifted off the axis of the Penning trap using patch potentials [2] and the magnetron frequency is extracted by varying the time these targets are allowed to orbit the trap center before being imaged with an MCP detector. The technique is faster than ECR as many target plasmas can exist in the trap at once without risk of heating with microwave pulses but is less precise, offering a magnetic field measurement with precision at the 1 Gauss level. The technique also relies on a calibration to ECR data for the first determination of the electric field at an axial position before the magnetic field can be extracted from later measurements.

The second term, where B_{Res0} is the remaining field contribution in ECR measurements but not captured by the magnetic field model in the initial well and B_{Res1} is the equivalent for the final well, brings the model into agreement with all ECR data taken in the initial and final wells at the 0.1 Gauss level. There is still some discrepancy between the stepped ECR data and the "magnetronmetry" results of around 0.9 Gauss.

Overall, the model's is accuracy is sufficient for the so called "Up versus down" style experiment described in Chap. 6. The field model is used in the particle pusher simulation that is run to simulate the entire experimental protocol. Antihydrogen are initialized and confined below the energy threshold for the trap, they are then pushed through magnetic fields as the mirror coil magnetic fields are lowered. The time and location of each antihydrogen escape and subsequent annihilation are recorded. In an effort to show any effect on the population as antihydrogen are accumulated during stacking (described in Chap. 2), this was also simulated as part of constraining simulation errors. The errors in this simulation have already been discussed in Sect. 6.4.2.

7.4.2 Summary of Bias Uncertainties

All of the above effects are summarised in Table 7.1 as shown in the recently published article [1]. The following section will be a brief summary of how the effects are calculated. These uncertainties are shown in Chap. 6 as a contribution on the bias axis of the release study data set (Fig. 6.7).

ECR spectrum width error: the average full width at half the maximum of Gaussian fits to the peaks in each ECR measurement in the initial and final field, with the mirror A and G measurements added in quadrature. The full width half maximum was taken as a conservative estimate of the error on the magnetic field as the electrons sample a range of magnetic fields which cause the broadening of the ECR peak.

Repeatability of ($B_G - B_A$): the standard deviation of all valid measurements, this excludes data that were corrupted by incorrect microwave settings etc as described in the high field measurement Sect. 7.2.1.

Peak field size and z-location fit: the residuals of the peak map quadratic fits in Sect. 7.3.1.

Field decay asymmetry (A to G) after ramp: using the data in Section 7.3.2 which tracked the fast decaying persistent fields, with the key parameter being a difference in the decay of the induced fields between the two locations. The largest measured difference was 0.1 Gauss, corresponding to the 0.02 g found in the summary table.

Bias variation in time: the standard deviation of the bias over the ramp down but weighted by the annihilation events. This magnetic field ramp profile was calculated using the magnetic field model after adjustments to fit the data.

Field modelling error: arises from the use of the magnetic model in the calculation of the bias variation in time. The error is quantified using the stepped ramp ECR results in Sect. 7.3.3. This is not the same field model error as in Sect. 6.4 which contains the impact of the simulation of the release experiment not the bias.

7.5 Future Perspectives

This section aims to show the first steps made towards new ECR techniques that can full characterise the magnetic fields required for a 1% determination of the gravitational acceleration of antihydrogen. The idea is to show ECR can be used in a dynamic field to characterise the magnetic field as a function of time and show the magnetic release is both controlled and reproducible.

7.5.1 Barriers to a 1% Measurement

First it is helpful to provide some context for the limiting factors that need to be overcome for a successful 1% precision determination of $a_{\bar{g}}$. The ECR campaign presented so far in this chapter has been highly successful and reached the required precision for this first determination of $a_{\bar{g}}$. It is clear the 1% measurement cannot be conducted without a significant ECR campaign requiring electrodes in the precision region of ALPHA-g. The design of such Penning trap electrodes for this purpose has been discussed in Sect. 4.1.4 and has overcome many technical challenges.

The proposed measurement in the precision region uses significantly lower magnetic fields, and by extension lower magnetic field gradients, which improved the precision of the existing ECR technique compared to the high field in the release experiments so far conducted. The magnetic field ramp profile, as well as being lower, is a slower ramp down to increase the sensitivity to the gravitational potential difference across the trap. As the ramp is slower and over a smaller range of magnetic fields this makes the prospect of ECR in a dynamic field more plausible. The next sections will show the first development studies undertaken with the aim of exciting the cyclotron resonance of target plasmas in a changing magnetic field.

7.5.2 Single Frequency ECR

In an attempt to show it is possible to resolve a difference in arrival time at a set magnetic field in ALPHA-g a measurement of the onset of heating is show in Fig. 7.16. The concept is that by holding target plasmas simultaneously under mirror A and G, then irradiating with the same microwave pulse as previously discussed locks in magnetic field information at a specific time. I set the microwave frequency to be

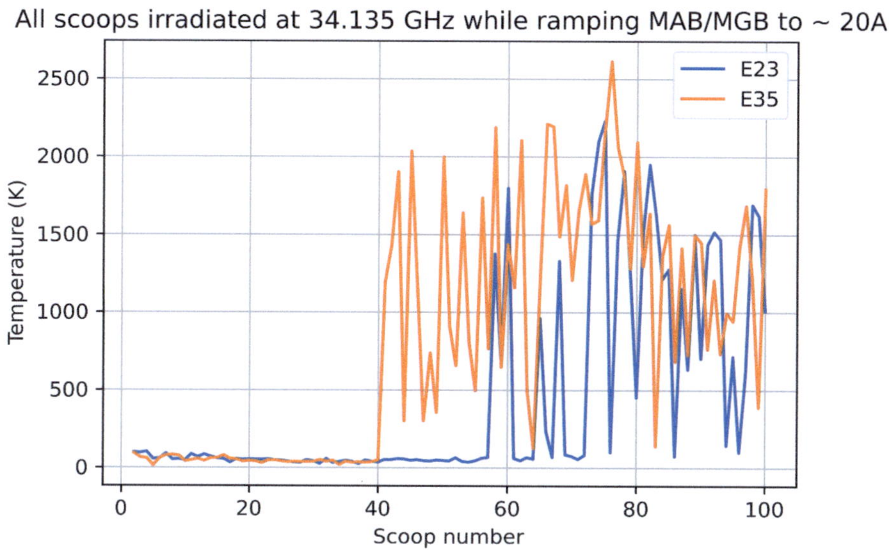

Fig. 7.16 Proof of principle measurement for observing heating onset during a ramp to a point

resonant with the end magnetic field after a ramp and then measures the temperature of many targets, all irradiated with the same frequency. The ramp is initiated and over time one should see an onset of heating as resonance is reached (as the ramp ends). If there is a delay between reaching resonance in each location this will be visible as a time (target plasma number) offset in when heating begins. A time difference was intentionally added to the ramp in Fig. 7.16 to show this is possible as a concept. The resolution is limited by the time between target extraction and the microwave irradiation but the first indication is that we have the capacity to observe differences in the field ramping on the order of 200 ms.

7.5.3 ECR in a Dynamic Field

The previous subsection then naturally leads to the question whether one can observe heating of a target plasma as the cyclotron resonance passes through a microwave sweep during a magnet ramp as in Fig. 7.17. This figure shows a linearly changing magnetic field in red and an example set of microwave frequencies in green. An attempt was made at this measurement and showed one could observe heating of a target plasma at a single frequency during the sweep, indicating that the cyclotron resonance has passed by within a time constrained by the on/off resonance targets.

In an attempt to improve the measurement and constrain more of the magnetic ramp profile a second study was undertaken. The procedure consists of holding electrons under both top and bottom mirrors and making five microwave sweeps

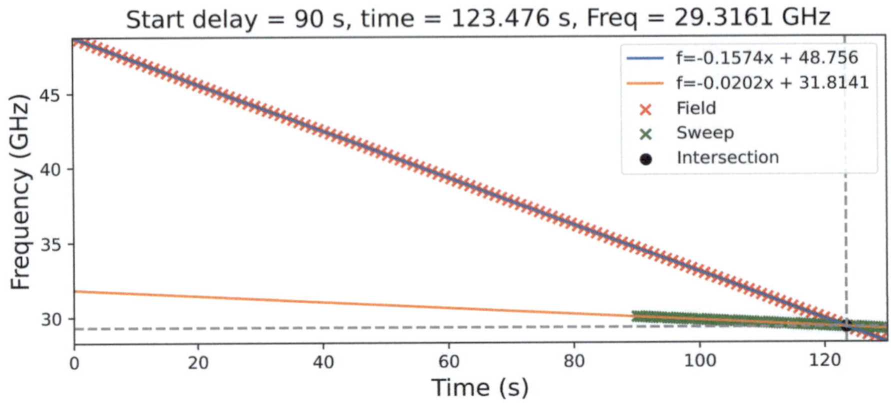

Fig. 7.17 First calculation for interception the magnetic field during a ramp

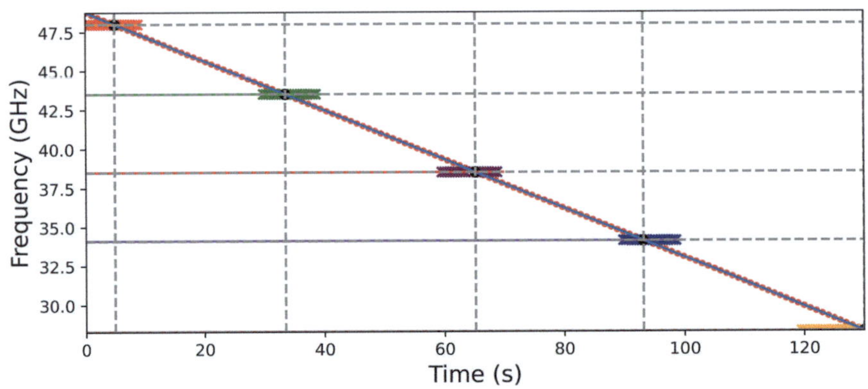

Fig. 7.18 Proposed intercept points for ECR in a dynamic field

during a ramp down, each intended to intercept the magnetic field at a set time as in Fig. 7.18. This proof of principle was carried out during a 130 s ramp in the "0g" magnet configuration so the fields should be identical at each location at all times. Heating of both electron plasmas was observed in all five frequency ranges at the expected time, indicating the field is the same at both locations and the rate of magnetic field change was correctly anticipated.

While this attempt showed it is possible to heat electrons at pre-determined frequency step it has major caveats. Most notably that while we see heating it is not determined in a single measurement if this is due to the ECR peak or a side band. Secondly, during the 10 ms microwave pulse used the magnetic field has changed by ~0.5 Gauss. I am satisfied that this proves we have the capability to perform ECR in a fast changing field in the future it is ultimately unnecessary for the "up versus down" classification experiment. This work is not at an advanced level and requires significant development but will become essential for higher precision gravity measurements in ALPHA-g.

7.5.4 *Conclusions of Work Towards 1%*

The studies shown here are the important first of many steps towards the characterisation of dynamic fields in ALPHA-g using ECR. The proof of concepts were carried out in large magnetic fields compared to those that will be used in future high precision measurements. Using lower fields improves these measurements considerably as the resonances is narrowed, the build up of persistent fields is significantly reduced and the range of microwave frequencies required reduces the chance of missing a resonance due the microwave power being low.

References

1. Anderson EK et al (2023) Observation of the effect of gravity on the motion of antimatter. Nature 621:716–722. ISSN:1476-4687. https://doi.org/10.1038/s41586-023-06527-1
2. Baker CJ et al (2024) Measurements of Penning-Malmberg trap patch potentials and associated performance degradation. Phys Rev Res 6:L012008. https://doi.org/10.1103/PhysRevResearch.6.L012008. https://link.aps.org/doi/10.1103/PhysRevResearch.6.L012008

Chapter 8
Conclusions and Outlook

I will take this opportunity to summarise the work that has been shown through this thesis, the work I believe needs to be completed in order to improve the precision of the determination of the gravitational acceleration of antihydrogen to the 1% level, and some final thoughts on the future for gravity experiments with antihydrogen beyond this precision.

8.1 Summary and Conclusions

The development of a new method to apply even tension to the ALPHA Penning traps (Sect. 4.1.1 has allowed the longest operation of a Penning trap at ALPHA without electrical shorts. These developments have also opened avenues for longer and more complicated Penning traps as shown with the latest design for the ALPHA-g top Penning trap in Sect. 4.1.4.

Microwaves have been successfully propagated into ALPHA-g at a broad range of frequencies (Sect. 4.2). This has allowed electron cyclotron resonance (ECR) magnetometry experiments that have been critical to the first measurement of the effect of Earth's gravity on antihydrogen. The hardware designs and implementation have also opened an opportunity for ALPHA to perform experiments in a much larger range of magnetic fields, which can improve systematic studies of antihydrogen spectroscopy as well as future gravity experiments.

ECR magnetometry has been shown as a vital tool for ALPHA in the development and characterisation of systematic uncertainties in gravity experiments. The addition of measurements in multiple locations simultaneously and rapid repetition of measurements has allowed the characterization of fast decaying magnetic fields and the spatial dependence of these decays. A major development in this thesis is the ability to make measurements in a high magnetic field gradient. This has enabled the co-location of magnets and annihilation detectors, as well as providing constraints for magnetic field profiles for antihydrogen motion simulations.

A. M. W. Powell, *Magnetic Field Characterisation for Gravitational Free Fall Measurements of Antihydrogen in the ALPHA-g Experiment*, Springer Theses,
https://doi.org/10.1007/978-3-032-07190-3_8

All of these developments have culminated in the first observation of the motion of antihydrogen under the influence of Earth's gravity. The initial aim of the campaign was to determine a sign of the gravitational acceleration of antihydrogen but with a successful experimental period and an effective magnetic control and magnetrometry campaign, it was possible to determine $a_{\bar{g}}$ was found to be 0.75 ± 0.13(statistical + systematic) $\pm\, 0.16$(simulation)g, where g is the gravitational acceleration (g $\simeq 9.81$ m/s).

This is the first direct measurement of gravity acting on antimatter and is consistent with no difference between antimatter and matter gravity within uncertainties. This measurement was significantly limited by the low number of antihydrogen available but rules out no gravity or repulsive gravity acting on antimatter. This is a landmark result that has built on decades of technological developments in antihydrogen research [1–4]. This first result provides a basis for higher precision measurements in the future to extract a value for $a_{\bar{g}}$ and further constrain any possible differences between matter and antimatter.

8.2 Towards a 1% Measurement

The future goal of ALPHA-g is to measure the value of $a_{\bar{g}}$ at a precision of 1 %, a target set given there is no compelling observation or theory to rule out a difference between matter and antimatter gravity at that level [5]. While there are several challenges from a magnetic field point of view, there are other considerations to be made. Firstly, the number of antihydrogen atoms used per release experiment must be increased. This has impacts on the background rates as fewer trials need to be completed to acquire statistics. The vertex reconstruction and background rejection must improve; this will allow for longer release times and increase the sensitivity of the experiment. Cooling of antihydrogen must be implemented; the colder the antihydrogen, the deeper in the magnetic well they sit and are therefore more sensitive to the measurement technique. This cooling could come in the form of laser cooling and adiabatic expansion cooling.

For a significantly higher precision measurement than the one claimed in this work, the full ALPHA-g apparatus would need to be employed. This includes the precision trapping region that has been designed to reduce the magnitude and possible asymmetry of any induced magnetic fields. Operation in this region is also at a significantly lower magnetic field and as such field gradients under the release coils are also substantially reduced. The implication for magnetometry is that ECR measurements can be made at a higher precision than for the work in this thesis. Considerable time and effort have already been spent to design, build, and install a Penning trap for the purpose of ECR in the precision region. The slower magnetic field change in the precision region also makes the prospect of ECR in a dynamic field easier to implement and make meaningful determinations of the magnetic field during a release experiment.

8.3 Beyond the 1% Precision Level

To reach beyond the precision level ALPHA-g was designed to measure is an interesting question and ppm level precision using trapped antihydrogen has been discussed previously [6]. This would require atom interfereometry which is also of interest for high precision spectroscopic comparisons of hydrogen and antihydrogen.

The HAICU (Hydrogen Antihydrogen Infrastructure at Canadian Universities) experiment has recently been approved at TRIUMF and is under development with an aim to produce a atomic fountain interferometer that is compatible with both hydrogen and antihydrogen [7].

References

1. Andresen GB et al (2010) Trapped antihydrogen. Nature 468:673–676. ISSN: 1476-4687. https://doi.org/10.1038/nature09610
2. Ahmadi M et al (2017) Antihydrogen accumulation for fundamental symmetry tests. Nat Commun 8(681):1–6. ISSN: 2041-1723. https://doi.org/10.1038/s41467-017-00760-9
3. Charman AE (2013) Description and first application of a new technique to measure the gravitational mass of antihydrogen. Nat Commun 4(1785):1–9. ISSN: 2041-1723. https://doi.org/10.1038/ncomms2787
4. Amole C et al (2012) Resonant quantum transitions in trapped antihydrogen atoms. Nature 483:439–443. ISSN: 1476-4687. https://doi.org/10.1038/nature10942
5. Fischler M, Lykken J, Roberts T (2008) Direct observation limits on antimatter gravitation. arXiv:0808.3929 [hep-th]
6. Hamilton P et al (2014) Antimatter interferometry for gravity measurements. Phys Rev Lett 112:121102. https://doi.org/10.1103/PhysRevLett.112.121102
7. Fujiwara M, Capra A, Alpha/Haicu Collaboration Team (2023) Fundamental symmetry tests via precision hydrogen-antihydrogen comparisons: ALPHA at CERN and HAICU at TRIUMF. APS Meeting Abstracts. APS Meeting Abstracts, p D14.006

MIX
Papier aus verantwortungsvollen Quellen
Paper from responsible sources
FSC® C105338

If you have any concerns about our products,
you can contact us on
ProductSafety@springernature.com

In case Publisher is established outside the EU,
the EU authorized representative is:
Springer Nature Customer Service Center GmbH
Europaplatz 3, 69115 Heidelberg, Germany

Printed by Libri Plureos GmbH
in Hamburg, Germany